the wildlife photographs

the wildlife photographs

JOHN G. MITCHELL

NATIONAL GEOGRAPHIC

WASHINGTON, D.C.

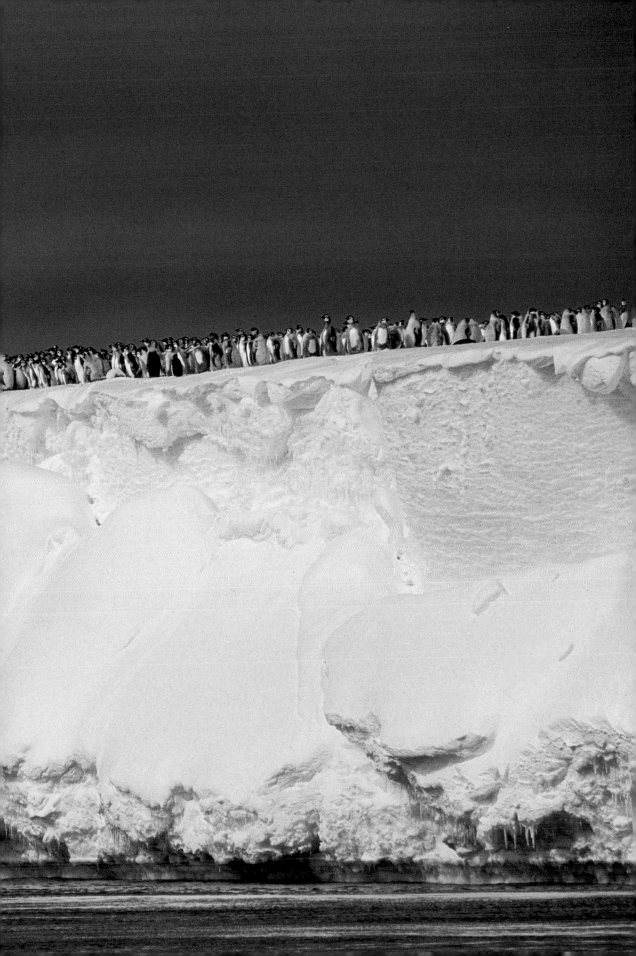

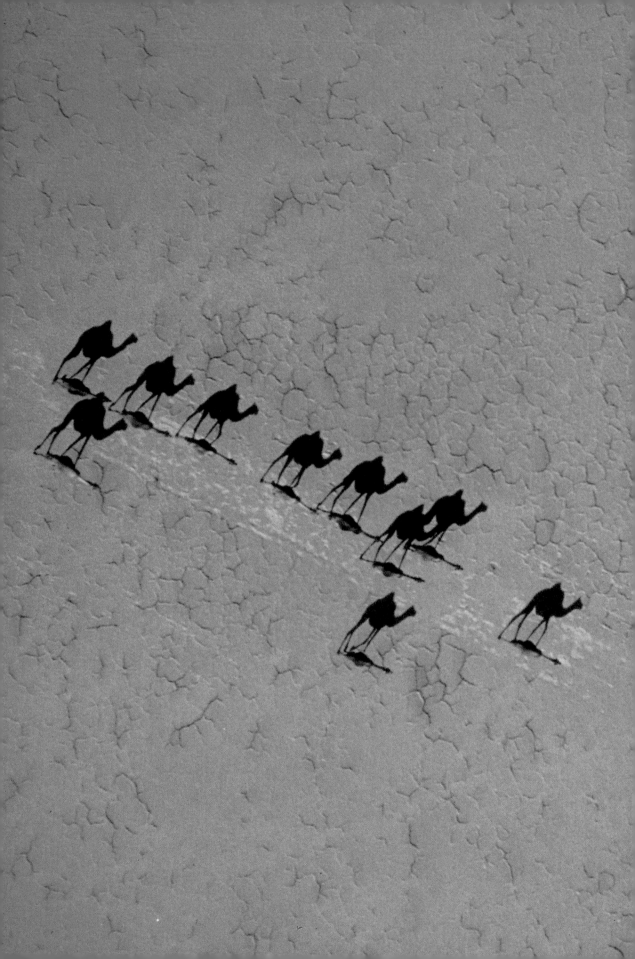

contents

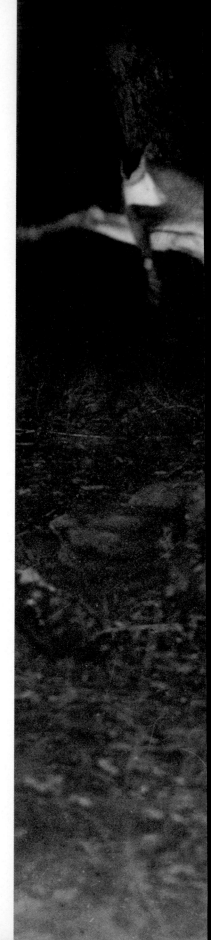

introduction

Nineteen-aught-three was a good year for getting things started. Down at Kitty Hawk, North Carolina, the Wright brothers defied gravity with newfangled wings and said hello to the Age of Flight. In New York City, a gentleman in a mud-splattered duster announced that he had just driven his horseless carriage all the way from San Francisco—first time ever—in only 69 days. And in Detroit, the Ford Motor Company sold its first automobile. Hello to the reign of King Car. And hello to the first wildlife photographs ever published in the journal of the 15-year-old National Geographic Society of Washington, D.C.: black-and-white pictures of Alaskan reindeer. In those days, of course, there weren't many photographs of anything in National Geographic, or in any other magazine. But it wouldn't be long before publication of trend-setting images of wildlife in wild places became one of the Geographic's strongest suits, and arguably its most popular.

Among the earliest trendsetters was George Shiras III, an avid outdoorsman, one-term Congressman, and pioneer in nighttime nature photography. Shiras, prowling the shore of Michigan's Upper Peninsula in a small boat, armed himself with a pan of flash powder, a box camera mounted on a bow swivel, and a collection of Rube Goldberg–type rigs

Startled white-tailed deer bolt through a Michigan forest in one of National Geographic wildlife photographer George Shiras III's pioneering trip-wire flash photographs.
GEORGE SHIRAS III.

designed to send nocturnal critters into flight or startle them in a freeze as the shutter snapped.

After a decade of trial and error (including one flash-powder explosion that forced George to abandon ship), Shiras was sufficiently confident of the results to show Gilbert H. Grosvenor, editor of the GEOGRAPHIC, a box of his plates. Grosvenor snapped up more than 70 photographs—of deer, moose, porcupines, raccoons—and spread them through 56 pages of the July 1906 issue of the magazine.

Those pictures, Grosvenor later recalled, "aroused tremendous interest in natural history." Not everyone was as interested as the editor, though. Two Society board members complained that "wandering off into nature is not geography," and they promptly resigned.

Over the years since, NATIONAL GEOGRAPHIC editors and photographers have repeatedly demonstrated that nature is a large part of geography, even as George Shiras may have been the first hunter to demonstrate that the camera can be more powerful than the gun. In the introduction to his two-volume book, *Hunting Wild Life with Camera and Flashlight*, published by the National Geographic Society, Shiras wrote of his satisfaction with

> the place the camera now occupies in the equipment of the sportsman as well as that of the field naturalist. The latter recognized its value almost as soon as dry plates were available. But the former was not convinced of its superiority to the rifle until improved apparatus enabled him to obtain photographs of big game which were not only more beautiful than mounted heads or rugs, but were far more impressive evidences of his prowess as a hunter.

Shiras, whom many consider the father of wildlife photography, may have been a shade too sanguine, implying that beautiful pictures might soon replace trophy heads on the walls of the sportsman's den. Well into the 1920s, NATIONAL GEOGRAPHIC's photographic take on large wild animals—and in those days, only the large ones qualified for editorial space—was pretty much oriented toward the celebrity safari. Elephants, rhinos, and lions tended more often than not to appear in the magazine

Geographers, biologists, and ecologists would divide the Earth into many more distinguishable environmental niches, but those divisions can be simplified and resolved into five regions: four major habitats plus the oceans.

in a permanently supine position, sometimes with a presidential pith helmet nearby. (Between March 1909 and January 1911, the magazine published four hunting articles by Theodore Roosevelt.) "Those were the days," says Annie Griffiths Belt, a gifted photographer in her own right, "when wildlife was viewed as a trophy, a conquest, or a meal."

The way Belt reconstructs it, the Society's wildlife coverage over the years evolved through a number of stages. After the "safari" period came the "gee whiz, the world is a zoo" era, during which animal species were pictured as attractive curiosities. Then, in the 1950s and 1960s, as the Society began to feature the conservation efforts of scientists and naturalists in the field, its photographers turned their collective lens on some of the devastating impacts human activity was having on wildlife around the world. Finally, says Belt, "the photographers' deepening respect for wildlife and concern for its future" led to coverage that would view wild species "in the context of their natural habitats and ecological interdependencies."

So what makes National Geographic's approach to wildlife photography not only distinctive in its field but also a good part of the reason its practitioners consistently walk away with the profession's top prizes?

Kent Kobersteen, Director of Photography at the magazine in the 1990s, believes the essential ingredient is realism. "We try for images that are real and not the result of staged situations," he says. "It's not enough to show what an animal looks like. We go to great expense—in both time and money—to send our natural history photographers into the field to portray natural behavior.

Kathy Moran, a picture editor who oversees many of the magazine's wildlife assignments, says "behavior is everything—and interpreting it correctly. That's the gift the GEOGRAPHIC can give a photographer. The time to sit there and wait for the real behavior to unfold before you. That's what it all boils down to. Portraits of an animal might be beautiful to behold, but by themselves they can't tell a story."

Asked to name the most important attribute a wildlife photographer—or any photographer, for that matter—can bring to NATIONAL GEOGRAPHIC, Kent Kobersteen doesn't hesitate with his answer: intellect.

George Shiras III and an unidentified paddler navigate Michigan waters. A former member of Congress and an outdoor enthusiast, Shiras took up wildlife photography in the late 1800s. This self-portrait exemplifies his innovative photographic techniques.
GEORGE SHIRAS III.

"The ability to put together aesthetically strong images is a given," he says. "But the depth of intellect necessary for a photographer to do a successful coverage of any type for this magazine cannot be overstated—especially when you're talking about wildlife coverage." Generally, Kobersteen says, the photographers are out there "working alone with very little direction from the office. They must not only have a thorough knowledge of their subject. They must have a passion for it."

Kathy Moran adds: "Our photographers have such heart and soul and commitment to the animals. I truly believe that they do more with their photographs to raise public awareness in support of conservation efforts than decades of work by the most dedicated scientists."

But how far can—or should—photography go in making a point or telling a story to stimulate public concern for wildlife? "These are dicey times for photographers," Peter Benchley wrote in an essay for NATIONAL GEOGRAPHIC, celebrating the career of underwater photographer David Doubilet, "times of enormous fundamental change in the way images are made and even in what images are." This is an age, he went on, "when the technology exists to create—actually fabricate—what once we were able to accept as reliable reflections of reality. Nowadays the cow can be seen to jump over the moon; a snail can be seen to swallow a star. . . . David will have none of it."

The name of the game is digital manipulation. "Photofakery" in an "Age of Falsification" is how the writer Ken Brower described it in a 1998 article in *The Atlantic Monthly*. "More and more digitally doctored images are appearing in the media," warned Brower, who has written or edited more than a dozen large-format volumes of nature photography. "The trend alarms a number of photographers. It worries certain editors and it worries me."

Though the trend cited by Brower involved the manipulation of wildlife photos in books or periodicals outside the National Geographic Society, the writer stated that the use of a computer to alter an image first emerged as an issue in 1982—right on the cover of NATIONAL GEOGRAPHIC magazine. To accommodate both the magazine's logo and the verticality of a horizontal image, a photo doctor had moved three camels in relation to the Pyramids at Giza. The adjustment was duly, and dourly, noted.

"That was more than 20 years ago," Kent Kobersteen says. "Yes, unfortunately, we did it then. Would we do it now? No." In fact, Kobersteen says he would not hire any photographer who used digital enhancement in display prints for advertising and rigorously eschews that practice in journalistic or editorial work. "My view is that this is a thin line on a slippery slope," says Kobersteen. "It calls into question the integrity of all of that photographer's work, and I just won't hire him."

There are other questions of what is and isn't acceptable in photographic reportage of wildlife, and there are wildlife issues where the lines are not quite so thin, the slope not so slippery, and the viewpoints of editors like Kobersteen considerably less absolute. Baiting raises just such a question. Though the presence of bait doesn't necessarily alter some predators' natural behavior, its use is considered by some to be bordering on creating an artificial situation. Among underwater photographers, baiting sharks to the camera is a common practice. "Is that right or wrong?" asks Kobersteen. "In most cases, that does not bother me. We do it. But it may bother someone else. There are no absolutes on things like that, only an infinite number of shades of gray."

On baiting, there seem to be fewer shades of gray when the animal is a mammal, especially a large predator. "I know you can't always draw a line in the sand," says Chris Johns, the magazine's Editor in Chief, among whose signature species are large African predators. "But I'm very opposed to the practice of baiting mammals. I don't believe any photograph is worth risking the life of an animal. And it would be a risk if that animal got habituated to handouts."

THE COLLECTIVE WORK OF THE PHOTOGRAPHERS represented in this book is global in its reach, eclectic in its selection, and egalitarian in its treatment of creatures both great and small. The images range from elephants in Central Africa to jellyfish in the Tasman Sea, from the long-horned beetle of French Guiana to the white wolf of northern Canada and the black bear of the United States. At first, the prospect of organizing such a congeries of critters seemed daunting. Sorting them geographically—a chapter, say, for each of the continents—was one obvious solution, but not very interesting. Taxonomy might have provided the organizing principle, with the species segregated in their respective phyla, Noah's Ark as Grand Hotel. But that approach risked luring the general reader into a state of abject narcolepsy. Fortunately, these flickering ideas were put aside early on by the book's editor, Leah Bendavid-Val, who wisely decided that a durable thematic focus for both text and photographs must be driven by habitat, by portraying wildlife behavior in the natural environments needed by each species to sustain its kind.

According to the World Wildlife Fund, the natural terrestrial world can be divided into 15 biomes, each defined by its distinctive mix of animals and plants but labeled for its dominant vegetation (or lack thereof). For this book we have reduced the number of general habitats to four, and tossed in the oceans for good measure. Temperate forests embrace two biomes, for example: coniferous and broadleaf. What we have elected to call "open country" embraces six biomes, from tropical savanna and temperate grassland to desert and dry shrub country (including such shrubby islands as the Galápagos). Then we have tropical forests (four different biomes), and the "ends of the Earth," the Arctic and the Antarctic (biomes of permanent ice, tundra, and taiga, also called boreal forest). There you have it—a structure that may or may not suit everyone. As for the animals, I suspect they wouldn't care a hoot, however we parse them.

The photographs in each chapter are bracketed by two sections of text: Up front, an essay on habitat; in back, a brief profile of a photographer and one of his signature species. In the essays we have tried not only to describe the physical habitats and prominent wildlife they sustain, but also to explore issues that are threatening the future of these ecosystems and animals. In each essay you will also hear about a few photographers who have not been given the profile treatment. People like Mark Moffett, a camera-toting ecologist so absorbed in the micro-world of tropical insects that he has been known to sit on the head of a poisonous snake and survive. Emory Kristof, aiming to be the first to record sperm whales preying on giant squid. Mattias Klum and Tim Laman, prowling the dipterocarp forests of Borneo. Jim Brandenburg, afield in the North Woods. Maria Stenzel, the photographer "with a penchant for high latitudes," perched at the edge of the Antarctic ice.

What this book does not attempt to be is a primer on photographic technique in the wild, an exposition of how certain images were created with certain kinds of equipment. For my own part, I wouldn't know an f-stop from a frying pan. What I do know, after nearly half a century in the business of making words, are wild country, lively landscapes, and some of the two-legged characters one is likely to encounter at the skinny edge or in the quaking guts of wild places. As it happens, some of the best of these characters are wildlife photographers. I am privileged to know a few of them, enough for me to understand why picture editor Kathy Moran salutes them for "such heart and soul and commitment to the animals."

I salute them, too. Having seen the splendid trendsetting, groundbreaking images these photographers are able to create, I have to confess that it just might be possible, after all, for one truly good picture to be worth much more than a thousand words. ➤

temperate forests

IT WAS A DIFFERENT WORLD WHEN THE HUMAN HAND first daubed a stroke of ocher or charcoal on the wall of a limestone cave. The glacial ice was in full retreat. In its place the warming land successively reformed itself under tundra sedge and taiga spruce, until a kind of open woodland emerged, with stunted broad-leaved deciduous trees beginning to poke their canopies among the evergreen conifers. The time? Who knows? We moderns call it Pleistocene. The place? Likely somewhere in Western Europe. The owner of the hand? Homo sapiens, Cro-Magnon. And what has Cro-Magnon rendered upon the inert limestone? A picture. It is a picture of an animal. There are pictures of many animals, primal images of reindeer and elk. Here is a mammoth, there is a bison—all these creatures that were hunted and killed by a people who could not only render the wild flesh into sustenance for their stomachs, but stylize their prey into forms sufficiently magical to stimulate the soul. To be sure, ocher and charcoal would evolve by and by into palette and brush, or film and camera. The dank wall of the cave would metamorphose into the fibrous page of a book. But still, this is where making pictures of wildlife really began, here at the ancient edge of a temperate forest that once covered most of the milder regions of the Earth.

Wandering through a Virginia forest, a black bear can't be distinguished from other bear species simply by color. Body shape tells the tale: It carries its rump higher than its shoulders and has a more convex face.
MICHAEL NICHOLS/NGS IMAGE COLLECTION.

23

And now edge is about all that is left of it. Of all the major biomes, none has been so thoroughly fragmented—in many places consumed—as the temperate forests of Europe, East Asia, and North America. Less than a third of this forest type remains within its potential range, which also embraces southeastern Australia, New Zealand, and southern Chile. In the eastern United States, many of the stands that pass for forest are simply regenerating on cut-over lands or abandoned farm fields.

The broad-leaved deciduous woods have taken the hardest hits. They lean toward accessible topography, which makes them more desirable as farmland or cityscape than the sometimes steep and difficult terrain of conifer country. Some scientists estimate that only one percent of the deciduous forest remaining in the Northern Hemisphere is true, scout's-honor, untouched, virginal old growth. The United States still harbors a few tiny pockets of ancient hardwoods here and there in the hollows of the Appalachians. Europe miraculously clings to the wild Bialowieza Forest on the border of Poland and Belarus. And in almost every temperate country, you will find a few "managed" woodlands, pruned and pampered by professional foresters. But elsewhere, the burnished colors of autumn foliage or the black tracery of leafless branches etched against a winter sky are sights seen mostly between the rooftops, or down the busy road and through the fenced fields to grandmother's house.

As habitat, temperate forests have served their wildlife well. Even so, though, some historic species of the temperate forest realms are no longer with us. Others crowd their names upon the lists of the threatened and endangered species of the world. And a few opportunistic ones in the United States—the white-tailed deer and the raccoon are good examples of this phenomenon—now fare better among gardens and garbage cans than they ever did in the unpeopled woods. Even the coyote, once a prairie critter, has taken to snatching mice and house pets from the shady groves of suburbia.

The larger North American predators—bears and cougars and wolves—have for the most part retreated to the wilder precincts of Canada and the protected parks and wilderness areas of the West, where photographers under contract to the National Geographic Society have been recording the animals' last best stand against the march of time.

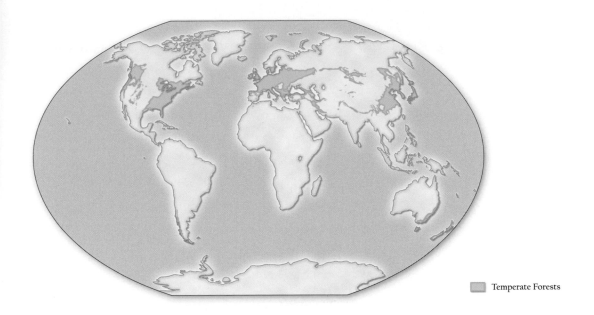

ONE OF THE GEOGRAPHIC'S MOST PROLIFIC photographers of North American wildlife subjects in recent years is Joel Sartore, a soft-spoken Nebraskan whose best work for the magazine has included stories about the future of the gray wolf, the survival of the grizzly bear, and the fate of lesser creatures protected under the Endangered Species Act. The last of these assignments resulted in Sartore's widely acclaimed book, *The Company We Keep: America's Endangered Species*, with text by Douglas Chadwick (National Geographic Society, 1996). (Photographs from that article and book appear in this chapter, and there is a profile of Sartore and "Grizz" beginning on page 48.)

In general, however, much contemporary photographic work in the temperate forest has little directly to do with wildlife, but quite a lot indirectly, insofar as the photographer's mission is to examine the condition of the habitat. Peter Essick, for example, is a photojournalist known not for his pictures of animals but for his coverage of tough conservation issues played out against the background of a threatened landscape—the U.S. National Forests, the Wilderness Preservation System, the Grand Canyon. Back in the States after an assignment in the northern forests of Canada, Essick spoke of the negative impact on wildlife of "hot and heavy" logging of birch and aspen to feed a big pulp mill in Alberta. "Migratory birds nest

Though few virgin stands remain, significant portions of three of Earth's continents—and slivers of two more—abound in either coniferous or broad-leaved temperate forests.

in those forests," he said. "But logging is chipping up their habitat." At the approach of the boreal winter, the migrants head south, many to the woodlands of the southern Appalachians. Essick lives near Atlanta, Georgia. There, he notes, the avian migrants face a loss of habitat no less disturbing than the clear-cuts of Alberta. Around Atlanta, urban sprawl is chewing up the Appalachian woods at a prodigious rate.

It happens that I am a native of this same eastern woodland region. The writer Rutherford Pratt liked to call it the "summer-green" forest, so I will, too. He described its original range as running the length of the Appalachian Mountains and spilling across them from the Atlantic Coast to the Mississippi River. By almost any definition, it was (and still is, patchily, though species vary by latitude and elevation) a forest of oak and ash, beech and maple, hickory and poplar, walnut and cherry, sour gum and sycamore, sassafras in the south and paper birch north. Not to mention the coniferous pines and cedars and hemlock, among many other arboreal genera.

Preying on mourning dove remains, an American burying beetle secretes an embalming fluid and applies it to flesh as it feeds. Decreasing carrion and increasing competition in the temperate forests endanger the insect.
JOEL SARTORE.

And all my summers were green, when I was growing up in the woods of southern Ohio. I do not recall seeing a great deal of wildlife other than the squirrels and rabbits I sometimes stalked with a .22 rifle snugged in the crook of my arm. Deer were almost unheard of those days, in those woods. Hunger from the Great Depression had turned the out-of-doors into a de facto commissary where jobless men endeavored to fill their cook pots with wild game. In some thickly populated regions of the Midwest and New England, it would be 20 years before a hiker or a hunter in the woods might see a possum or a porcupine, much less a deer. Perhaps that is why, in those years of our wildlife scarcity, my older brother felt obliged to invent a predator to keep things lively in the woods beside our house. He called it the mangamoonga. It was said to resemble a hairy rhinoceros, though not exactly, and to be able to slog through the mud without leaving a track. I never encountered the mangamoonga, but that was okay. It seemed more than enough just to imagine the great beast lurking there, somewhere deep in our otherwise stingy woods.

Woods need wildlife; it's not just the other way around. The ecologist Aldo Leopold, describing the landscape of his beloved North Woods, once observed that it all added up to "the land, plus a red maple, plus a ruffed grouse." He figured that in terms of physics, the grouse could represent only one-millionth of either the energy or the mass of a single acre. "Yet subtract the grouse," he wrote, "and the whole thing is dead."

However degraded the temperate forests of North America appear today, they are in far better shape than their counterparts elsewhere in the Northern Hemisphere, though Siberia still clings to substantial tracts of a wild transitional forest (a forest, technically speaking, that is more boreal than temperate). In contrast to North America, Europe and East Asia have endured many more centuries of expanding human settlement and all that settlement entails—land clearing for agriculture, timber harvesting for home construction, and the gathering of firewood for heating and cooking.

Consider China and the plight of the giant panda. Reports from Sichuan Province tell of a loss of prime forest habitat, attributed to rapid population growth in rural communities in and around China's largest panda sanctuary, the Wolong Nature Reserve. Pandas there favor relatively flat territory covered with mixed deciduous and evergreen trees as well as bamboo, the panda's meal ticket. Unfortunately for the endangered creature, this is also where people cut firewood.

And consider Europe. Consider its paucity of tree and wildlife species compared to North America's relative abundance. For diversity of tree species, North America outnumbers Europe five to one. But the imbalance is not altogether attributable to Europe's longer exposure to human activity. Europe was dealt a bad hand by the alignment of its mountain ranges.

The Alps and the Pyrenees—the Carpathians, too, if you want to stretch it a bit—run east and west. During the ice ages in Europe, these mountain ranges presented a barrier to plant and animal species as they were retreating south from the advancing glaciers. Consequently, some species became trapped and were terminated with extreme and frigid prejudice. In North America, on the other hand, the Appalachians, the Rockies, and the ranges of the Pacific Coast are more or less aligned in a north-south direction, thus allowing species the freedom not only to flee the ice, but to return north after the ice had retreated.

Botanists have parsed the North American temperate forests in so many elaborate ways that I am going to risk their wrath by oversimplifying the distinctions into three major types: the summer-green eastern woods, the western pinelands, and the temperate rain forests of the Pacific Coast. And, as might be expected, some of the photographers represented in this book have, in turn, parsed themselves into associations with one or another of these forests.

Let's start with summer-green—not the Appalachian part of it, but rather the part that hunkers around northern New England, southeastern Canada, and the Upper Great Lakes, the part that is popularly known as the North Woods. Canoe country, much of it. A land and waterscape redolent of voyageurs in stocking caps, of portages and traplines, wet paddles flashing in the sunset, the haunting cry of a loon.

In the November 1997 issue of NATIONAL GEOGRAPHIC, photographer Jim Brandenburg took the reader on an extraordinary journey into a piece of this very country. He was just outside the Boundary Waters Canoe Area Wilderness of Superior National Forest in northern Minnesota.

"In a dark spruce forest—two lakes and a portage from my bush camp—I have discovered a place of mystery and wonder." So begins Brandenburg's "North Woods Journal," an illustrated account of how, for the 90 days between the autumnal equinox and the winter solstice, he made but one photograph a day. "There would be no second exposure," he wrote, "no second chance. My work would be stripped to the bone and rely on whatever photographic and woods skills I have."

Stripped or not, the bone had substance to it. Though most of his 90 images are of North Woods landscapes and inanimate things, perhaps a third include wildlife—mallards, grouse, heron, eagle, deer, mink, coyote, wolf. He wrote: "All around me I witnessed cycles of life and death—with deer becoming wolves, bones becoming soil, lichens eating rocks, herons stalking fish." The experience rekindled "a deep primordial feeling, perhaps the same feeling an ancient hunter had, an emotion that I first experienced as a boy tracking foxes across the snow-covered prairie where I was raised." Brandenburg said that the project changed him. No doubt it changed many a reader's perceptions of the North Woods, too.

(There is yet another forest directly north of our North Woods, but one could hardly classify it as temperate. I call it the boreal Big Woods, for it sweeps unbroken from the granite ledges of Labrador to the butt end of Hudson Bay, then skedaddles in a widening curve to embrace those other grand lakes—Athabasca, Great Slave, and Great Bear—on its way to the Upper Yukon and the interior of Alaska. Four thousand miles across, spruce and fir almost all the way, the North American boreal forest forms the edge of one end of the Earth, and we shall visit it in our concluding chapter.)

The western pinelands of the United States and Canada are montane forests mostly, hunkering along the lengths of the Rocky Mountains and such other interior ranges to westward as the Sierra Nevada. These are the woods we associate with the scenic crown jewels of the continent—Yellowstone and Yosemite and Banff. Of pines there are a number of dominant species, site to site, latitude to latitude: towering ponderosas, lanky lodgepoles, massive sugar pines (which John Muir called "the very god of the woods"), ancient bristlecones (at 4,000 years old or more, oldest living things on Earth), gnarled pinyons of sun-burnt canyon country. Douglas firs outnumber the pines in many places, and in the Sierra, the giant sequoia still makes a stand in protected groves. Yet for all their impressive botanical majesty, these conifer forests would be zoological deserts were it not for the lowly deciduous shrubs and trees that bracket the regions' streams and bogs and wildfire burns. Aspen, birch, willow, and alder are more generous to wildlife than their evergreen neighbors. The conifers may provide cover, but the leaf-droppers render food and shelter—seeds for the birds, stems for the dam- and den-building beaver, browse for the elk and the moose. And where browsers roam, can predators be far behind?

TODAY IN THE CONTIGUOUS UNITED STATES, there is only one place big enough and wild enough to sustain all three of temperate North America's top-o'-the-food-chain predators—the mountain lion, the grizzly bear, and the gray wolf—and that is the greater Yellowstone ecosystem in the Rocky Mountains of Wyoming, Idaho, and Montana. These same precincts, however, weren't always so hospitable to big predators. Sheep and cattle interests in the northern Rockies had never looked kindly on brother wolf, and a war of extermination soon had the canid in full retreat across the Canadian border. But as the final years of the 20th century ticked by, a few scattered packs began skulking back into Montana, and the U.S. Fish and Wildlife Service decided to enhance their numbers by capturing wolves in Canada and releasing them into the wild of Yellowstone National Park.

NATIONAL GEOGRAPHIC assigned photographer Joel Sartore to cover the return of the native. "It was a great challenge," the Nebraskan recalls. "It tested my patience. After two weeks on the story, I didn't have a single picture. You could see some wolves, but they were too far away even for a telephoto lens. And there was no way you could get any closer without spooking them. So I called my editor, John Echave, and said, 'Hey, I'm wasting your money.' And John said, 'Stick it out and see what you can get.' Sure enough, a few days later four wolves came down off the mountain and from then on I was able to get pictures."

On the last day of his shoot in Yellowstone, Sartore encountered a pack of young wolves feeding on an elk carcass on the far side of a meadow. It was evening, hardly light enough for a picture, but Sartore decided to try. As he approached, the wolves stopped feeding, woofed at him, suddenly ran at him, and then started circling him about 30 yards out. "You know," Sartore says, "there is no record of a healthy wolf attacking a human in North America. I knew that at the time, but knowing didn't help. Sure, I was scared." Popping his flash in a futile effort to scare the pack off, Sartore backed slowly toward his vehicle—a tiny dot in the distance. But the wolves didn't scare; they followed. "I could hear them whining and panting in the dark, almost up to the door of my vehicle."

Later, some wolf researcher had a chuckle on hearing the photographer's tale. "You mean the pups almost got you?" he said to Sartore.

As it happened, those "pups" were young wolves, raised partly in captivity and thus habituated to humans. "I shouldn't have approached so close," Sartore says. "They were on a kill. They were doing what wolves do. But I'd hardly call 'em pups. Why, any single one of them was big enough to pull down an elk. And an elk's a whole lot bigger than I am."

FINALLY, THERE IS THAT RAIN-DRIPPING, FOG-SHEATHED FOREST at the edge of the Pacific Ocean. The lofty redwoods of northern California get it off to a good start, and so does that melting pot of many conifer species in the Klamath-Siskiyou highlands at the Oregon border. Northward along the shore of the sundown sea, across Washington State's Olympic Peninsula and British Columbia's Vancouver Island, the forest flexes its muscle within the massive trunks of Douglas fir and western red cedar, then scoots on to the fjords and archipelagoes of southeastern Alaska.

Here, in the United States' largest national forest, the Tongass, the monarchs among all towering trees tend to be the western hemlock and the Sitka spruce. Rutherford Platt—that literary chronicler of the American wilds and author of the classic books *Wilderness* and *Great American Forest*—once wrote that the crown of the Sitka spruce is often flat because it is formed "to spill the wind," and that "its ranches, which all point inland with the airflow, are rigid streamers twanging in the sea blast." Then, hardly missing a beat, Platt added: "It is appropriate that such rugged forest, with salmon-laden streams, should be the home of the world's largest carnivore, the Alaska brown bear, which weighs three-quarters of a ton and stands 12 feet when it rears up on its hind legs."

For my own part, no other woods have so powerfully impressed me as this Tongass forest. I haven't yet seen enough of it, or enough of

The bald eagle, symbol of America, mates for life. Both sexes achieve their distinctive coloring—white head and tail feathers, dark brown body—at about 5 years old. Many live as long as 28 years.
JOEL SARTORE.

the Earth's other great forests, to trot out that good-ol'-boy phrase and declare that this is as good as it gets. But then . . . who knows? Maybe this *is* as good as it gets.

I suspect I was thinking that way not so long ago when a float plane out of Juneau dropped me and a friend at a place called Windfall Harbor, on the shore of Admiralty Island, in the Kootznoowoo Wilderness of Tongass National Forest. We spent only two days on the island, tenting a sensible distance from the nearest salmon stream because a run of humpbacked pinks were spawning there. Coming in over the harbor, we had seen eagles and a big brown bear picking the spawners out of the gravelled shallows, and we didn't want that bear or any of its brethren picking us out of a flimsy tent. Of course a coastal brown bear is not likely to do something like that when there are so many succulent salmon about. Still, caution is the better part of valor in bear country. And Kootznoowoo is definitely that kind of country, said to support the highest density of brown bears in North America. In fact, the name Kootznoowoo is Tlingit for "Fortress of the Bears." Russian fur traders who passed this way in the 19th century had a different phrase for the place. They called it Ostrov Kutsnoi—Fear Island.

On that first day, we saw a few bears at ground level, at a safe distance across the harbor, and on the second we decided to hike up the beach to the mouth of the spawning stream. But when we got there, Grizz was gone. Not a predator in sight, not even an eagle. My friend figured the bear (or bears) were napping somewhere upstream, bloated with salmon. "We could stroll up there a way," he said. I looked at the narrow stream valley tapering into the rain forest between thick stands of alder. "You could," I said. "I'll watch." We turned away then and headed for camp.

I have often thought of that experience in that special place. And every time I do, I recall Aldo Leopold's equation about "the land, plus a red maple, plus a ruffed grouse" in his Wisconsin North Woods, though I put a Kootznoowoo kind of spin on it. I see a stream valley lined with alder, plus a flattop Sitka spruce, plus a salmon-fat grizzly bear, and I figure that, in terms of physics, the bear could represent only the slightest fraction of either the energy or mass of five acres of that forest. But subtract the bear? Subtract the bear and the whole thing is dead. ➴

Signs of a raven's landing in fresh,
powdery Vermont splendor
resemble a child's snow angel.
BERND HEINRICH.

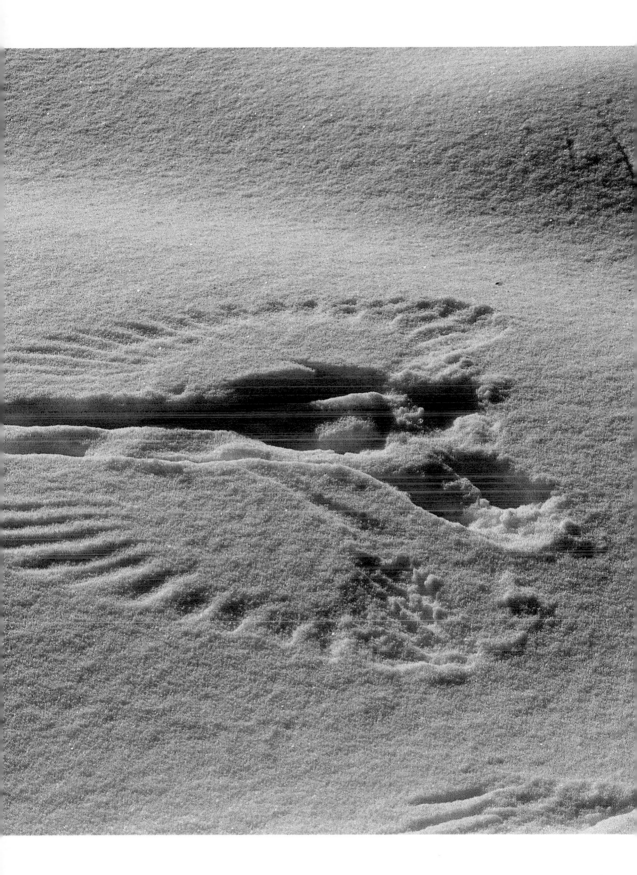

ABOVE: *Whistling swan glides on frigid waters, its spotlessly white feathers obscured by fog. Smaller and fainter in call than the trumpeter swan, the whistling swan nests farther north, in Arctic regions.*

JIM BRANDENBURG.

OPPOSITE: *Balancing on a tree limb, a giant panda takes a snooze in its temperate forest habitat in China. Hunting for high-priced pelts has threatened the species, but today poachers face 10 to 20 years in jail.*

LU ZHI/NGS IMAGE COLLECTION.

Residing in a preserve, a gray wolf protects his meal. Wolves eat about ten pounds of meat each day— preferably that of large hoofed mammals. But kills can be unpredictable, so these fierce canines often gorge on up to twice that amount when they can.
JOEL SARTORE.

Endangered Indiana bats hang together in Kentucky. Long cave networks in Indiana, Missouri, and Kentucky host thousands of these creatures as they hibernate in humid darkness through the winter.
MERLIN D. TUTTLE/BAT CONSERVATION INTERNATIONAL.

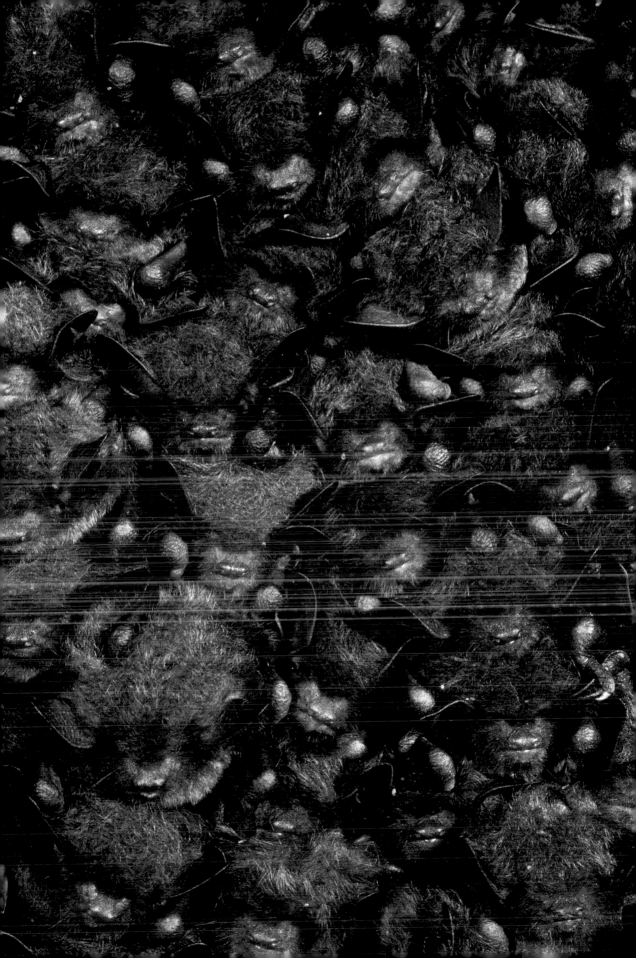

Paddling through clear temperate waters in Florida, this snapping turtle may weigh 200 pounds and live to be 100 years old.
GEORGE GRALL.

FOLLOWING PAGES: *Top dogs in the food chain during the long Minnesota winter, this shadowy gray wolf pack survives because its herbivorous prey, such as elk, weaken from malnourishment.*
JOEL SARTORE.

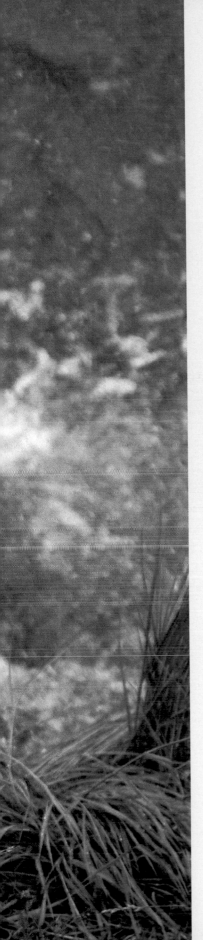

joel sartore &
grizzly bears

IF THERE IS ONE THING JOEL SARTORE WISHED people knew about grizzly bears, it would be this: "They're not the ferocious man-eaters that everybody thinks they are. They would prefer to run away from conflict." As it is, North America's largest predators evoke a wide set of responses, "from ooh-snoogie-woogums to God save us," as Douglas Chadwick wrote in a July 2001 article in NATIONAL GEOGRAPHIC magazine, a collaboration with Sartore. Helping us understand the bear was writer Chadwick's job. The task of getting us to see the bear more clearly was Sartore's.

The photographs here were all shot at Katmai National Park in July, a time of year when grizzlies find easy prey in salmon swimming upstream to spawn. "The bears are concentrating on building up a fat reserve for the upcoming winter," Sartore explains, "so they're gorging themselves on fish." Some ate so much during these salmon feasts that they came up out of the water, laid down, and napped on the bank.

Sartore and Chadwick spent 16 weeks following bear spoor through Yellowstone and Alaska, mostly by plane or boat. They would figure out the bears' patterns, then set up in an unobtrusive place from which to photograph them. To get the shot of the bear in the waterfall, Sartore used an observation platform that photographers often use, especially during

Alaskan grizzly makes landfall, slinging off excess water after diving in for a meal of cold river fish.
JOEL SARTORE.

PRECEDING PAGES: *Grizzly perches at a fall's precarious edge, pawing its prey of sockeye salmon, as a seagull swoops in for scraps.* JOEL SARTORE.

ABOVE: *Grizzly sow nurses twin cubs in Katmai National Park, Alaska. Taking close-up family photos like this can be risky business for the photographer.* JOEL SARTORE.

salmon spawning season, so bears ignore them—to the point, Sartore marvels, of "even nursing their young close to the platform."

While Sartore was capturing images, Chadwick managed to bust some misconceptions, including one that long held the large brown bear of Kodiak Island and the coast of Alaska to be a separate species from the smaller grizzly found inland and down through the northern Rocky Mountains. Not so. All brown bears are of the same genus and species, *Ursus arctos,* and are found in the higher latitudes of Eurasia as well as in North America. Most Americans call them all grizzlies.

Over weeks in the field, Sartore often felt threatened but was charged by a bear in a serious fashion only once. "A sow with two cubs," he says, "is a hair-trigger combination. I had come too close. She charged to within 15 feet and stopped short when I lowered my eyes and slowly backed off." Even though he has continued working with bears—including close-up work with trained grizzlies for a National Geographic Television Special—Joel Sartore has not been attacked by a bear since.

The likelihood of someone being attacked by a grizzly is miniscule, he says. "Most of the time, the bear just wants to get out of your way." According to Chadwick, throughout the 1990s grizzlies in North America injured an average of seven humans a year and killed an annual average of two. That's not a whole lot when you consider that there are an estimated 58,000 grizzlies, half of them in Alaska, and that humans are increasingly pushing into their habitat. "Every year," says Sartore, "there are more and more people and fewer and fewer bears. Historically, the grizzly was out on the plains, but we pushed it back into the mountains. Well, now we're settling the mountains too. So this is the bear's last stand."

Today in the United States outside Alaska, the grizzly bear is listed as a threatened species, with no more than an estimated 1,200 individuals scattered across five or six ecosystems. Some 400 to 600 are believed to inhabit the Greater Yellowstone region of Wyoming, Montana, and Idaho. Another 400 or so are in Glacier National Park and the national forests of the northern Rocky Mountain Front. The remainder live in small, isolated populations in Montana, Idaho, and Washington State. Meanwhile Canada's bears, roughly 25,000 in number, are dwindling because of hunting, aggressive policies against nuisance animals, and backcountry development—"the same combination," Doug Chadwick wrote, "that put the lower 48 grizzlies on the imperiled list."

"If we're going to have healthy populations in the lower states," says Sartore, "we've got to restore some bear-friendly corridors between the isolated pockets." A joint U.S.-Canadian network of 800 groups is now working to restore and maintain wilderness from Yellowstone to the Yukon that could support the bears. "A lot of people are working on it," says Sartore. "Whether it becomes a meaningful reality remains to be seen."

Like his colleagues in the field, Joel Sartore wants to make every story count. "And the grizzly bear counts in a big way," he says. "Why? This big, far-ranging creature needs wilderness to survive, and what habitat you save for it you save for every other creature that is smaller, including people."

After years of photographing them, Joel Sartore knows bears well. "Grizzly bears are very intelligent," he says. "All they need is enough suitable habitat and hopefully they'll be able to survive into the future." ⌒

open
country

Perhaps it would be a pardonable stretch to say that open country is where sky becomes the roof of the Earth while earth becomes the floor of the sky. Open country is desert and prairie, savanna and steppe, wetlands and scrub. It is where, to borrow from the imagination of the writer Wallace Stegner, one is likely to encounter a wind "with the smell of distance in it." Tundra and polar ice are open country, too, but that's a different part of our story. Here we aim to examine the grasslands and arid landscapes that provide habitat for so many of the world's wild creatures, and proving grounds for so many of the world's wildlife photographers.

There are half a dozen extensive desert biomes throughout the world—the Gobi of Central Asia (*gobi* meaning "waterless place" in Mongolian), the vast Empty Quarter of the Arabian Peninsula, the Simpson and Great Victoria Deserts of Australia, the Mojave and Sonoran and Chihuahuan Deserts of North America, and Africa's Sahara, at 3.5 million square miles the planet's largest. Each has a fauna well adapted to arid conditions. But the number of species to be found in true, waterless desert country is relatively low, as might be expected.

Moreover, unlike the Earth's temperate and tropical forests, many of these deserts are expanding rather than shrinking. Some observers claim

Wildebeests teem over a cliff in Tanzania, part of their 500- to 1,000-mile annual migration. Fording rivers and lakes in such large herds can cause injury and drowning for calves and weak swimmers.
MITSUAKI IWAGO.

that, on a global scale, human activity is converting brush and shrublands into desert at the rate of 40 square miles a day. That's a lot of desert. The worst of it is occurring in Africa's sub-Saharan Sahel region, where a burgeoning human population, pumped up in part by refugees from civil strife in Chad and Sudan, is robbing the land of its marginal vitality through overgrazing, intensive cultivation, and the stripping of brush for fuelwood.

Even without human interference, few deserts contain themselves within neat boundaries or, for that matter, fulfill textbook expectations of what a desert ought to look like. Consider, for example, the Kalahari in Africa, which covers most of Botswana and parts of South Africa and Namibia. Though the northern reaches of this great basinlike plain are often tagged geographically as part of the Kalahari, they do not qualify as desert climatically, for they receive more than ten inches of rain annually and are puckered here and there with open woodland and thorn brush interspersed with savanna. From the highlands of Angola, far to the northwest, tributary runoff finds its way to the Okavango River. The river in turn feeds a vast network of wetlands called—curiously, inasmuch as the nearest ocean is 500 miles away—the Okavango Delta. It is a region rich in wildlife: elephants, zebras, giraffes, buffalo, impala, lions, cheetahs, leopards, baboons, African wild dogs. It is the kind of lively open country that has drawn photographer Chris Johns—now Editor in Chief of NATIONAL GEOGRAPHIC—to Africa time and again.

Johns's first African assignment for the magazine was in 1988, taking him north of the Kalahari in the Great Rift Valley that runs down from Ethiopia into Kenya and Tanzania. It was also his first exposure to large wild animals, though images of East African cultures and landscapes would be included in his story as well.

"When you go to a place like Africa," Johns recalls, "you realize that the truth is a complicated matter. There are so many layers to a place like the Rift Valley. If you try to peel them away, you only discover how little you know." Ten years after the Rift assignment, Johns knew enough about African wildlife to land two NATIONAL GEOGRAPHIC cover story assignments—one spotlighting Botswana's wild dogs, the other on cheetahs, "Ghosts of the Grasslands" in the Okavango Delta. (You can

Open Country

read more about Johns and the cheetahs he photographs in the profile that starts on page 80).

Botswana's Chobe National Park is another brush-country location that has attracted a number of wildlife photographers and filmmakers over the years. But few could possibly know it as intimately as Dereck and Beverly Joubert, whose Africa films for the National Geographic Society have won numerous awards, including Emmys for such television specials as "Reflections on Elephants" and "Eternal Enemies: Lions and Hyenas," both filmed in Chobe. In their book, *The Africa Diaries: An Illustrated Memoir of Life in the Bush* (National Geographic Adventure Press, 2001), Dereck Joubert writes of a lion-hyena encounter on a damp evening in 1991. The filmmakers had come upon a pride of lions attacking a herd of wildebeest, when, Joubert notes in his diary entry of April 30, "suddenly the night was alive with the calls of hyenas. Hyena eyes bobbed and danced menacingly in every direction, too many for us to count." And then, later, "at the two lion kills, the hyenas were massing, drumming themselves up into a frenzy. . . . A lion ran in and charged them, changing the dynamics in favor of the lions at this kill, but leaving the second kill vulnerable. The hyenas switched without hesitation to swamp the lions at that kill. Lions scattered everywhere as hyenas viciously bit escaping lions."

Desert and prairie, savanna and steppe, wetlands and scrub—open country covers large portions of Africa, Asia, Australia, and much of southern South America. North America's open country, the prairies and plains, changed significantly beginning with the European arrival.

Southwest of the Chobe and the Okavango Delta, the Kalahari takes on more of a desert look, though scattered waterholes and drought-tolerant shrubs and grasses manage to sustain a variety of species, including kudu, eland, and springbok. Farther west, the Kalahari reaches toward the Namib Desert on Namibia's Skeleton Coast, a shoreline speckled with the exposed ribs of wrecked ships.

The Namib is said to be one of the driest deserts in the world. Pinched between the Kaokoveld highlands and the sea, it is a place of large dunes, few oases, and little rain—an average of half an inch a year. But where the Namib hurts for rain, it wallows in condensation. Offshore, the frigid Benguela Current, sweeping north from the Antarctic, meets warm, moist Atlantic air. The result: fog. At night, inshore, the fog condenses into droplets of water on foliage and the surface of rocks. And where there is water there has to be life. In the Namib, life takes many forms, from the termite to the elephant.

Some time ago, with a base camp by the ocean, the Australian filmmakers Des and Jen Bartlett worked along the Skeleton Coast, photographing elephants and lions and other creatures adapted to the Namib's harsh environment. Though a 25-mile-wide strip along the shore enjoys official protection as Namibia's Skeleton Coast Park, the Bartletts discovered that poachers had taken a heavy toll on the elephant herd in the decade before their arrival. To locate survivors, the couple scouted oases from the air in two-seater lightweight aircraft called Drifters. Over the course of their eight years in this region, the Bartletts saw evidence that the elephants were making a comeback.

The lions of the Skeleton Coast, however, were not as fortunate. "When we first arrived," the Bartletts wrote in NATIONAL GEOGRAPHIC, "at least a dozen lions lived there. Now there are none." The problem was livestock. In the highlands east of the Skeleton Coast, herdsmen kept their cattle and goats along dry riverbeds where groundwater often wells up in springs. When prey near the coast became scarce, the lions would head for the springs. And when the lions attempted to prey on the livestock, the herdsmen would shoot them. "Saddened by the loss," the Bartletts wrote, "we hope that in the future this tragedy can be replaced by something positive, both for the local inhabitants and the wildlife."

GRASSLANDS TEND TO BE MORE FORGIVING of wildlife, although predator-livestock conflicts are hardly unknown in the feral pastures of the world. Among the most extensive are the savannas of East Africa, the great Eurasian steppe that sweeps west from Manchuria almost to Ukraine, the short-grass plains of Western Australia, and South American grasslands such as the Mato Grosso in Brazil, the Llanos of Venezuela and Colombia, the Gran Chaco in Paraguay, and the Pampas of Argentina. In North America, there used to be a place called the tall-grass prairie; now it is known as "the corn belt." And west of all that corn lie the Great Plains, their greatness gravely compromised by decades of irrigated farming and insufficient regulation of grazing on government lands.

Considered together, before they were tamed, the American prairie and plains probably formed the greatest of all the Earth's grasslands, reaching one thousand miles from the edge of the Great Lakes and the lower Ohio Valley to the Rocky Mountains, and from Canada's Saskatchewan River almost to the Rio Grande. In the eastern precincts of this region, the prairie was dominated by a grass called big bluestem, which grew to heights of five to nine feet and sustained itself through a system of roots so dense that, end to end, one square yard of sod could yield 20 miles of capillary fiber. You'd think a grass that tall, sprouting from a mat that thick, might defy any effort to tame it. But think again. Think about that innovative blacksmith named John Deere out there in mid-19th-century Illinois at the edge of the tallgrass, tinkering with steel. Voilà! The all-steel, one-piece moldboard plow. Follow its furrows through the 20th century, and you'll find that 90 percent of the tallgrass prairie—nearly half a million square miles of it—is gone.

An eminent historian of the American West, Walter Prescott Webb, once opined that the steel plow was part of a trinity of contraptions that finally wiped the wild out of the prairie-plains, the other two being the

Cape gannets, large South African seabirds, nuzzle affectionately—a ritual they repeat every time a mate returns to the nest.
CHRIS JOHNS.

59

Colt revolver and the barbed wire fence. The way Webb told it, the plow destroyed the tallgrass, the wire forced cattle in confinement to trample the shortgrass, and the revolver permitted a galloping horseman to get off six shots without missing a stride. Usually, the man with the Colt was a sodbuster, a cattleman, or a soldier. And who was at the deadly other end of the gunsight? Usually an Indian. I'd add one more contraption to Webb's West-taming trinity and make it a quartet. I'd throw in the Sharps rifle, the weapon that came to be known as "the buffalo gun."

One cannot mourn the loss of the North American prairies without remembering and mourning as well the wild ruminant herds that once thundered across these lands in numbers beyond counting. No one will ever know how many buffalo were out there before North American sunlight first glinted on European armor. The best guess is 60 million. (And, to be taxonomically correct, we should stop calling the American bison a buffalo. Buffalo belong in Asia and Africa. Bison belong in North America and northern Europe, where a related species is known as the wisent.)

By whatever name, though, the big shaggy herbivores of the prairie-plains shaped an entire aboriginal culture for centuries, especially after the Plains Indians acquired the Spanish horse. The bison was the western Native American staff of life. Even so, what the tribes took from the herds was negligible until blue-eyed rovers arrived from the East, looking to trade for hides. By 1850, the market was claiming nearly a quarter million bison hides a year.

Then the railroads pushed out across the plains. William F. Cody signed on with the Kansas-Pacific to supply the construction crews with fresh meat. In a year and a half, "Buffalo Bill" is said to have killed 4,000 bison, all from horseback.

But dropping the animals for their tongues and hump ribs and leaving the rest to buzzards and coyotes, wasn't the half of it. In 1871, tanners discovered that the hide of a bison, heretofore used as a robe, could be turned into fine leather. That and a national recession two years later loosed upon the plains a horde of hungry men with trigger itch. Most of the hunters scratched their fingers on the trigger of a Sharps, which had an effective range of a thousand yards or more and caused much understandable awe among the tribes. Some Indians called it the gun that "shoots today, kills tomorrow."

The near extermination of the bison also contributed hugely to the extermination of the Plains Indian way of life. In 1875, General Philip Sheridan, the same of Civil and Indian Wars fame, announced that the hide hunters had done "more to settle the vexed Indian question than the entire Regular Army has done in the last 30 years. They are destroying the Indians' commissary." The general went on: "Let them kill, skin, and sell until the buffaloes are exterminated. Then your prairies can be covered with speckled cattle and the festive cowboy, who follows the hunter as the second forerunner of an advanced civilization." Old feisty Phil. When it came to understanding what was civilized and what wasn't, he didn't have a clue.

SOME SAY IT'S A MIRACLE that the American bison managed to elude the terrible fate of the passenger pigeon, bouncing back from the brink of extinction in the 20th century. More likely it was no miracle at all but rather the result of a captive breeding program at the Bronx Zoo at the turn of the century, belated protection imposed by the U.S. Department of the Interior, and the efforts of a handful of western ranchers who believed that consumers would soon develop an appetite for "buffalo" meat. The ranchers were right. Today, more than a quarter million bison graze across private ranches and public lands, mostly in the West, converting grass into the protein of tomorrow's steaks and burgers.

Anyone who thinks that ranching has tamed the bison as decisively as the plow once tamed the prairie should talk to the photographer Sarah Leen. She once decided she wanted a dramatic close-up of a thundering herd at roundup time in Colorado's Rocky Mountain Bison Ranch. To get it, Leen had to hide her camera in a box, under sagebrush, and then, as the bison galloped by, trip the shutter with a buried cable. The setup allowed Leen to stay out of harm's way, but it was fraught with peril for her equipment. "Those bison banged up a couple of sets before we got what we wanted," Leen recalls. "But we lost only one of our cameras. Fortunately, it was an old one."

There's wild open-country spirit in those bison yet.

Jim Brandenburg is another photographer with more than a passing respect for bison. Brandenburg, who now lives in the North Woods of Minnesota, grew up in the far southwest corner of that state, a hop and jump from the South Dakota line and a skip short of some remnant ghosts of the tallgrass prairie. Portraying the tallgrass, and the effort to establish a Tallgrass Prairie National Park, was the first of his many assignments for NATIONAL GEOGRAPHIC.

"If you're not careful," Brandenburg says, "you can get hooked on the prairie. I wasn't careful. It has an amazing power to draw you in. Now it's deep in my blood"—so deep that he founded the Brandenburg Prairie Foundation, which, together with the U.S. Fish and Wildlife Service, has bought more than 800 prairie acres in Rock County, Minnesota, not far from the photographer's birthplace. I asked him how a patch of prairie that big escaped John Deere's moldboard plow. "Simple," said Brandenburg. "Too many rocks in it."

Pair of dung beetles in South Africa roll animal excrement into a ball that they will soon bury. Returning later, they will either snack on it or use it as a soft and tasty nest for eggs and hatchlings. CHRIS JOHNS.

I suspect that open country of one kind or another gets into the blood of most photographers who care as deeply about the natural world as Jim Brandenburg does. Take Joel Sartore, for example. Forests, mountains, and rocky shores have yielded much of his best work with wildlife—yet Sartore is a son of the Nebraska prairie, and so fiercely proud to be one that he helped set up a grassroots organization called the Conservation Alliance of the Great Plains. The group is pushing for creation of a system of national prairie preserves on the northern Great Plains. "Wouldn't it be grand," says Sartore, "to have a place where you could see a free-roaming herd of bison and prairie dog towns? And no fences."

But Sartore doesn't confine his advocacy efforts to the Cornhusker State. In South Texas, not far from Houston, he discovered a species at the brink of extinction—Attwater's prairie chicken. Only about 50 of the birds remained in one last stronghold, surrounded by urban sprawl and oil refineries. The Nature Conservancy was attempting to acquire additional lands nearby to create a sustainable breeding habitat for the birds. Sartore saw a story and contacted the writer Doug Chadwick, with whom he had already collaborated on a number of stories involving endangered species.

"The situation isn't as attention-grabbing as the grizzly's plight," Chadwick noted in response to the photographer's story proposal. "But it may be a clearer measure of our relationship with native wildlife. The solution could hardly be more straightforward: The creature needs a little more room to live. . . . We're not talking about tearing down dams or retooling a whole industry. . . . We just have to figure out whether anyone will come up with a minor real estate adjustment before this life-form expires from neglect." Sartore's proposal was approved.

Over the years, protection of open-country habitat has benefited from the work of many photographers, especially in the Everglades of South Florida and the red rock canyons of Utah. I'm thinking in particular of the photographers who, with books and articles, pioneered coverage of these places when they were most at risk and thereby exposed their values to a wider audience.

For the Everglades, let's scroll back more than 30 years to a time when that great "river of grass" (Marjorie Stoneman Douglas's phrase) and the national park through which it flows were staggering under the burden of a severe drought. Levees and canals constructed by the Army Corps of Engineers had already blocked much of the overland flow of water from Lake Okeechobee to Florida Bay. Real estate developers were attempting to drain the cypress swamps west of the Glades. And the Dade County Port Authority was busy advancing a plan to build the world's largest jetport only six miles north of Everglades National Park. When critics complained, the boosters of flood control, drainage, and development simply responded, "What's more important? Alligators or people?"

Several photographers at work in South Florida believed that the place ought to be big enough for both. Among them was Frederick Kent Truslow, whose specialty happened not to be gators but birds—such Floridian birds as the limpkins and wood storks that he would feature in the pages of NATIONAL GEOGRAPHIC. And there was Patricia Caulfield, who spent the better part of five years in and around the Everglades in the late 1960s and whose photographs, as much as any others, helped block the jetport, introduced GEOGRAPHIC readers to the Big Cypress National Preserve, and bought this wild realm some time.

In an endnote to her 1970 book, *Everglades*, Caulfield expressed some of the same sentiments I would hear decades later from Jim Brandenburg, describing how open country can get in one's blood. "Loving the Everglades came naturally to me," Caulfield wrote, "for I grew up in the flat plains and prairies of Iowa. Superficially, the saw grass river resembled my native habitat. I felt at home. But the initial visual impression did not last for long. I soon glimpsed a wildness I had always missed, without knowing it, in the tamed farmlands of my childhood."

For Utah, it's another 30 years of scroll-back to a time when the mining industry was knocking at the gates of the Escalante wilderness and "motorized mass industrial tourism," in the words of Edward Abbey, was threatening to skewer Canyonlands National Park with asphalt highways. And whom do I see there? I see Philip Hyde, protégé of Ansel Adams and perfectionist with the four-by-five view camera, planting his tripod atop a sandstone needle in a Canyonlands place called Chesler Park.

I can hear Hyde, too. I hear his words tumbling out of *Slickrock*—the book he co-authored with Abbey in 1971—his comments on the Escalante:

> In the marvelous acoustics of these vaulted chambers the canyon wren's melodious, descending trill flows inward, making the hearer as delighted as a bird to be there listening to the wild music. Walking days are filled with stream crossings. . . . If you are lucky, you may come upon a recently implanted set of cougar tracks on a wet sandbar.

Looking shocked but trying to look shocking, a captive nine-banded armadillo in Florida springs into action. Its defensive repertoire also includes the ability to roll up into an impenetrable ball.
BIANCA LAVIES.

FOLLOWING PAGES: *White pelicans course serenely around a river curve in the Mississippi Delta. Flocks can number up to 2,000.*
ANNIE GRIFFITHS BELT.

We all are lucky that Philip Hyde listened to the music of the wren and followed the track of the cat on the sandbars of the Escalante. In large measure because of Hyde—and the many writers and photographers who would follow his tracks across the red rock canyon country—those "vaulted chambers" now lie ensconced within Grand Staircase–Escalante National Monument. As for Canyonlands National Park, it remains unskewered—though a few barbarians will always be barking at its gates. ➴

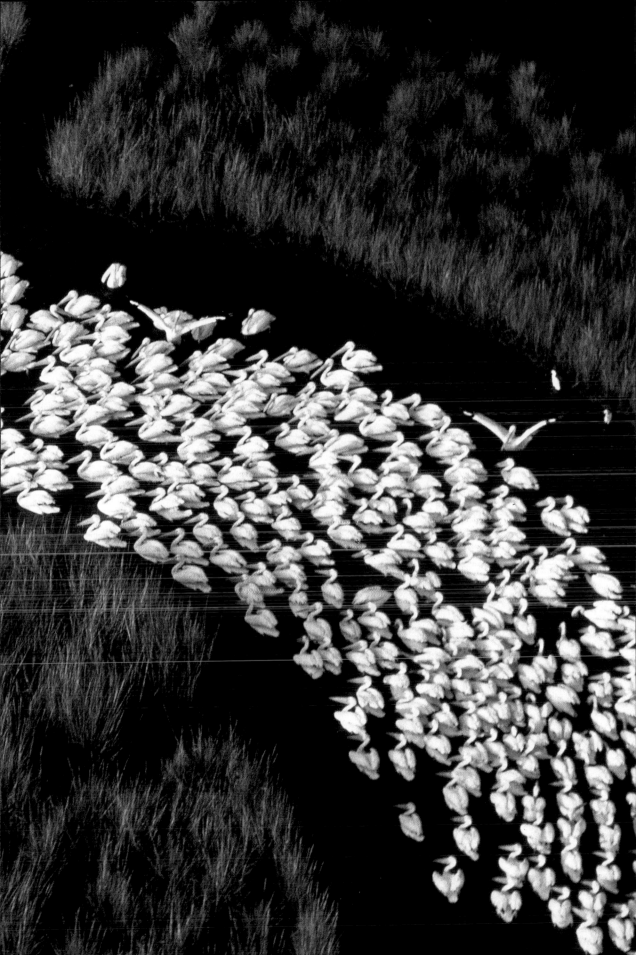

Cheetah and cubs feast on a fallen impala in the Okavango Delta in Botswana. These cats will be on the move shortly—their best defense. Most cubs fall victim to hyenas, lions, or other predators before reaching adulthood.
CHRIS JOHNS.

FOLLOWING PAGES: *Jackass penguins, named for their bray, live only on the south coast of Africa. Guano harvests in the 19th century robbed them of nesting material and threatened the species, but their numbers are now rebuilding.*
CHRIS JOHNS.

Eastern diamondback rattlesnake drapes its muscular body on exposed mangrove roots in the Florida half-light. In this pose, it is not dangerous; the rattlesnake must coil before it strikes.
CHRIS JOHNS.

Camera captures a placid giraffe wandering through the South African mist. The species may tower as tall as 19 feet and run at speeds up to 35 miles per hour. CHRIS JOHNS.

The skin of a white rhinoceros—
actually a shade of grey—glistens
in the grasslands sunlight of
Hluhluwe-Umfolozi Park in South
Africa as this bull emerges from a
cooling mud bath.
CHRIS JOHNS.

FOLLOWING PAGES: *African wild
dogs roam Botswana's grasslands.
Traveling up to 20 miles a day, they
search for prey to kill and
disembowel. These beasts are the
last of a dying breed: In 2004, only
about 5,000 remained.*
CHRIS JOHNS.

chris johns
& cheetahs

IN JANUARY 2005, CHRIS JOHNS WAS NAMED the sixth Editor in Chief of NATIONAL GEOGRAPHIC magazine. He was a GEOGRAPHIC staff photographer for years before that, shooting everything from cliffs above Hawaii's white-sand beaches to squirrels on the White House lawn. Roughly a third of the stories he has shot for the magazine have been about Africa. His work has also appeared in two National Geographic books, *Valley of Life: Africa's Great Rift* and *Wild at Heart: Man and Beast in Southern Africa.* Of all his assignments over the years, Johns figures the most challenging was his back-to-back coverage of African wild dogs and cheetahs in the Okavango Delta region of Botswana. "Both animals are remarkable predators," he says, refusing to choose a favorite. Readership surveys clearly tip the hat toward the cheetah, as might be expected of one of Earth's most charismatic animals.

Posted to the Okavango with the writer Richard Conniff, Johns found that the cheetah, unlike the leopard, or the wild dog for that matter, is very choosy about its food. "This is a very prey-specific predator," Johns says. "In the Okavango, they go after impala mostly. They'll hunt along a rim of woodland, run the impala down, kill it by suffocation, biting down on the windpipe. Then they'll take the kill into the brush." Cheetahs are not often successful in protecting their kills from other animals. "They lose a lot to

Cheetah drinks from shallow water while its companion looks suspiciously toward the camera. The photographer's challenge is to convince a wild animal to ignore him.
CHRIS JOHNS.

hyenas and lions, sometimes even to vultures. For all that speed and killing power, they're really such delicate creatures."

Throughout the cheetah's shrinking range, the gravest threat is the spread of pastoral agriculture, which not only destroys habitat but also presents livestock herders with a preemptive rationale for killing predators. "In Namibia," says Johns, "as many as 8,000 cheetahs were killed by farmers in the 1980s, before any conservation efforts could be put in place."

Long extirpated from the Asian continent, cheetahs may once have ranged even farther from their African homeland. "There's some evidence," the photographer notes, "that cheetahs may have been present in North America as recently as 12,000 years ago. Why is it that the pronghorn can run 60 miles an hour? There is no reason in historic times why this animal needs to run that fast, but 12,000 years ago, the pronghorn may have had a reason—the cheetah. And 60 miles an hour is roughly what the cheetah can do."

Cheetahs are known for hunting and killing their own prey. They prefer fresh kill and teach cubs to hunt for themselves at an age as young as six months old. Despite their speed, when pursuing prey such as the fast-darting gazelle, they succeed only about half of the time. When they do catch their prey, often by tripping them or pulling them down, cheetahs will employ a deadly lock on the victim's throat, to cause suffocation—a necessary ploy, since cheetah teeth are not long enough to kill prey by impaling them. Then these lithe cats must eat quickly before hyenas or other large beasts move in to steal their food.

Like many of the GEOGRAPHIC's most prolific wildlife photographers, Chris Johns did not embrace this specialized field when he first went to work with a camera. An Oregonian, Johns apprenticed on daily newspapers, doing what he calls "people stuff." He was working for the *Topeka Capital-Journal* when he got his first assignment from NATIONAL GEOGRAPHIC, tailing firefighting crews across the West. Johns then bombarded the magazine's photo chief with one story idea after another. "I think there were 17," he says. "And every last one was rejected." Meanwhile, he kept hearing the imagined voice of an early mentor. *When are you going to get serious about your pictures?* the voice wanted to know.

No voice, imagined or otherwise, asks that question of Chris Johns now. He has turned from photographing cheetahs and picked right up with African wild dogs. "I think that was the hardest story I ever tackled," he says. "I kept seeing these great pictures that I couldn't take. I'm in a Land Rover doing 45 miles an hour, trying to hang on, and I see these dogs just do astonishing things—jumping over brush six feet high, twisting and turning in the air. Wild dogs in pursuit of their prey can make even cheetahs look clumsy."

Cattle farmers have waged relentless war on these dogs for decades. Pup mortality runs high, mostly from predation by lions and hyenas. "These dogs are the most endangered large predator in Africa," Johns says. "The trouble is, they don't fit. Herders hate them and other people think they're goofy and ugly-looking. I tend to think they're incredibly beautiful."

Because the African wild dog is not widely regarded as a charismatic critter, Johns knows that his pictures of them are not, as he puts it, "going to win any kudos in the photographic community." But that's okay, so long as the pictures help people understand what's at stake in the ecosystem. "If the dogs can't survive in the Okavango Delta," Chris Johns believes, "then we really better rethink what we're doing to this planet." ᴖ

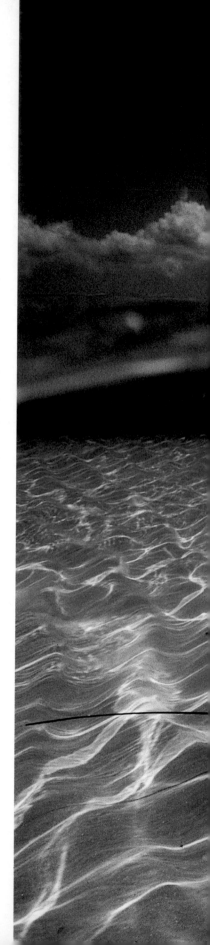

oceans

IN JANUARY 1927, NATIONAL GEOGRAPHIC published the first color photographs ever taken beneath the surface of the sea. They showed a brown-striped hogfish, yellow-eyed snappers, and blue-striped French grunts cruising in tropical waters off the Dry Tortugas in Florida. This historic achievement, a collaboration between Charles Martin, director of the Society's photo laboratory, and W. H. Longley, a diving ichthyologist, charted a new course for the GEOGRAPHIC. Thereafter its photographers would be probing the deep to bring us, often in the most startling and vivid colors, images of the multitudinous creatures of the seven seas.

If one were to highlight the Society's most adventurous and innovative underwater cameramen of the last half-century, one had best start with Luis Marden, who, over the course of a long and far-flung career, wrote and photographed some 60 stories for NATIONAL GEOGRAPHIC— though not all of them required Marden to get wet. That particular condition first affected the photographer in an important way in the mid-1950s, when he plunged into the Indian Ocean and the Red Sea with the filmmaker Jacques-Yves Cousteau.

"The luminous transparency of the warm water bathed me in light," Marden wrote in his story, "Camera Under the Sea," the magazine's first

Southern stingrays ride the ripples in shallow waters off Grand Cayman Island. Related to sharks, stingrays dig for crustaceans, fish, and worms. The spiny barbs on their tails prong predators.
DAVID DOUBILET.

full-length illustrated underwater feature, published in 1956. "I seemed to hang suspended," he went on, "in the heart of an enormous liquid sapphire." No doubt others have experienced similar fantasies when working in this blue world that covers almost three-quarters of the Earth. Bill Curtsinger, swimming with gray reef sharks in Bikini Lagoon or cruising with penguins in Antarctica's McMurdo Sound (more about Curtsinger later, starting on page 114). David Doubilet, tailing giant manta rays off the Yap Islands in the western Pacific or jellyfish in Monterey Bay. Flip Nicklin, a man once called "as relentless as Ahab" in his friendlier pursuit of all manner of whales (read about Nicklin's polar photos, starting on page 181). Emory Kristof, a specialist in deepwater biology and an innovator of advanced equipment to probe the secrets of the sea.

There is one fantasy, however, that no one is likely to experience nowadays, and that is the view, once widely held, that the oceans are too strong to be fractured, too big to be bruised. Marden himself saw that grand illusion dissolve before his eyes.

As for Cousteau's awakening, Marden would write of that turnover in a tribute to his old friend, "Master of the Deep," in the February 1998 issue of NATIONAL GEOGRAPHIC. First, Marden recalled the "halcyon days when the undersea world was new and lay all before us, waiting to be discovered. Every dive was like a visit to another planet." But as the years passed, Marden continued, Cousteau "began to notice something disquieting. In many places fish were growing scarce, formerly crystalline waters were increasingly murky, and once richly carpeted bottom now lay bare. . . . Appalled and angered, Cousteau the diver and filmmaker became Cousteau the environmentalist."

Today there is much about the seven seas to appall and anger all of us as we try to juggle the beautiful and lively images illustrating this chapter against the sobering facts of the oceans' biological degradation. Overfishing, destruction of marine habitats, toxic pollution—each has struck a heavy blow. Now some scientists are predicting that global warming could deliver the coup de grâce to many marine species already approaching the brink of extinction.

Coastal zones—the estuaries, salt marshes, mangrove forests, tidal flats, and seagrass meadows that cluster along the landward

edge—are among the world's most productive maritime ecosystems. And, arguably, the most imperiled. Some three-quarters of all commercial fish and shellfish landings in the United States, for example, are of estuarine-dependent species. Yet according to the nonprofit SeaWeb Project, despite recent efforts to enhance protection, the United States continues to lose coastal wetlands to waterfront development at the annual rate of about 30 square miles, or what the Project estimates is an area one and a half times the size of Manhattan. The Chesapeake Bay is said to have lost 90 percent of its seagrass meadows since 1900. Worldwide, mangrove forests survive along only half the coastlines they historically buffered. So it goes.

The United Nations Food and Agriculture Organization reports that commercial fish species in more than half of the world's major fishing regions are now in such serious decline that the stocks may be unable to regenerate themselves. Cod, haddock, flounder, and halibut numbers off the northeastern United States have hit historic lows. Newfoundland has shut down all ocean fishing—and with it tens of thousands of jobs. Bycatch—the random landing of nontarget species with longlines, gill nets, and other indiscriminate gear—takes a frightful toll of sea turtles, seabirds, dolphins, and porpoises. The Gulf

From frigid polar waters to tepid tropical bays, oceans cover nearly three-quarters of Earth's surface, presenting the broadest, deepest, and most daunting habitat of all for wildlife photographers.

of Mexico shrimp trawling industry has routinely caught, killed, and discarded four pounds of juvenile fish for every pound of shrimp retained. Atlantic swordfish hit the skids in part because of longline bycatch; 40,000 undersized "swords" were inadvertently hooked in 1996 alone. The slow-to-mature bluefin tuna, prized in the sushi and sashimi market, is likewise in trouble.

So are sharks. According to SeaWeb, a growing demand for shark fin soup in China and other Pacific Rim countries has increased shark landings in U.S. waters dramatically. In the United States, SeaWeb reports, "a pound of dried shark fins can fetch $200; in contrast, the rest of the shark can fetch as little as 60 cents a pound (wet weight). Consequently, in many places, fishermen simply cut off the fins and dump the sharks overboard." Their studies show that "the U.S. alone exported 575,000 pounds of shark fins to Hong Kong in 1988, and the market has continued to increase."

And as the fish come out of the water, the pollutants pour in. Pesticides, PCBs, and dioxins. Sewage effluents and industrial wastes. Airborne fallout from incinerators and power plants. Nutrient overloads from farms and feedlots. Eroded sediments from clear-cut logging. Urban stormwater runoff laced with heavy metals and lawn fertilizers. Oily bilge from boats and marinas.

So it goes? Yes. But it's still such a beautiful blue world when you only look at the pictures.

NO PHOTOGRAPHER CURRENTLY ACTIVE in that other world has done more to show us the ocean's beauty—even as he warns us of its perils—than David Doubilet. A longtime friend and sometime collaborator, the writer Peter Benchley, has described Doubilet as "a master of the intellectually challenging, physically taxing, always changing, and utterly arcane world of underwater photography." Not content with that high praise, Benchley added: "I warrant that in his field, David has no betters and few, if any, peers."

A native New Yorker, Doubilet began snorkeling at the age of eight, at about the same time, more or less, that Luis Marden was with Jacques Cousteau, testing the waters . Four years later, the youngster was scuba diving with a Brownie Hawkeye sheathed in a rubber bag. In 1972 he shot his first story for NATIONAL GEOGRAPHIC: "The Red Sea's Gardens of Eels." In the years since, he has produced more than 50 stories for the magazine, in addition to several books.

From an oeuvre covering so many subjects and so many seas—inasmuch as Doubilet has dipped into almost all of them—a landlubber such as I, who has dipped into none, is hardly qualified to judge which pieces show the master at his best. But I suspect you'll find a few of his best on the following pages. Take that Tasmanian saw shark photographed in Australian waters (page 97), for example—a monstrous conglomeration of vulnerably soft underbelly, probing mouth and nostrils, and brittle, spiny snout. Or that juvenile saltwater crocodile rising for a sniff of air in Australia's Jardine River (pages 104-105). Of that assignment, Doubilet recalled: "To get some of the shots I wanted, I had to get so close that, for a time, the camera was sitting where the croc's food should have been. At the last possible minute, I'd yank the camera out of there. I guess that's the real meaning of 'bait and switch.'"

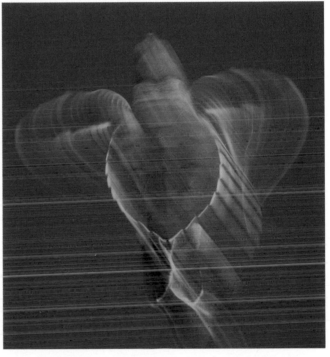

Black turtle shimmies through blue depths off the western coast of Mexico. Inhabiting the eastern Pacific, including the Galápagos, fewer than 10,000 of these turtles may survive.
BILL CURTSINGER.

Doubilet guided GEOGRAPHIC readers in 1999 to a "Coral Eden" in the western Pacific (one of those photographs can be seen on pages 102-103). But once there, among the reefs, he found that these "oases of life" were being plundered by "blast" fishermen using the kind of fertilizer-and-diesel-oil explosives favored by terrorists, while other buccaneers were pouring cyanide over the reefs to stun fish and then collect them for the aquarium industry.

Doubilet's most popular story in recent years, however, was the one that snagged the cover of NATIONAL GEOGRAPHIC in April 2000, after he teamed up with Peter Benchley to shed some sympathetic light on the great white shark—a quarter century after Benchley had demonized the creature in the book and film, *Jaws*. "I couldn't possibly write *Jaws* today,"

Benchley confessed. "Today we know that these most wonderful of natural-born killers, these exquisite creatures of evolution, are not only not villains, they are victims in danger of—if not extinction quite yet— serious, perhaps even catastrophic, decline." To obtain some of his remarkable close-ups of the real jaws of a feeding white, Doubilet used a camera hidden in a "seal cam," a rig built by photoengineer Walter Boggs to replicate the silhouette of a floating seal.

CHARLES "FLIP" NICKLIN IS ANOTHER PHOTOGRAPHER who has spent the better part of his career in the water with large critters, notably dolphins, porpoises, and whales. Known to his colleagues as "Whale Man," Nicklin, like Doubilet, got his start at an early age, flippering into the Pacific with his father, Chuck, an underwater cinematographer. In the introduction to one of Nicklin's books, *With the Whales* (text by James Darling), the cetacean researcher Kenneth Norris described the photographer as an "intrepid explorer of the blue domain" who has "played with these foreign creatures in a kind of underwater tag, where he was the 'little one from land' swimming among the behemoths, each regarding the other with the most surprising gentility."

Over the years, Nicklin has contributed some 20 articles to NATIONAL GEOGRAPHIC, about half of them profiling the behavior—and the gentleness—of whales and dolphins. The blue whale (largest of all the Earth's creatures). The killer whale. Three arctic species, the narwhal, the beluga, and the bowhead. The bottlenose dolphin. The sperm whale, Herman Melville's Moby Dick. The humpback whale. The minke whale.

The only whale I have ever been close to, though hardly as close as Nicklin, was a bowhead, dead on the beach at Barrow, Alaska. It was about a year after Nicklin's bowhead article had appeared in the GEOGRAPHIC, much of it illustrating the Inupiat Eskimo's legal but quota-controlled hunt for the whale in the Beaufort Sea.

My whale-on-the-beach was the third to be taken by hunters that season. Men stood atop it with long-handled cutting tools, flensing (or stripping off) chunks of black skin and pink blubber, the savory and nutritious muktuk that would be distributed among Barrow's Eskimo families. In an earlier season, one of the hunters, speaking of the Inupiat culture, had told Nicklin that muktuk is "the lifeblood of my people." And I suppose it is if, generation after generation, ancestral centuries of eating it mean anything.

But doubtless there was a time before quotas when, no fault of the Eskimos, muktuk was scarce. Historically, the blubber of one big adult bowhead, at 60 feet of length and 80 tons of heft, was said to be capable of rendering 6,000 gallons of oil. In the 19th century that kind of oil helped fuel the lamps of Europe and America. And the bowhead's flexible baleen proved to be the perfect material for manufacturing umbrella ribs and corset stays.

So the whalers came to the Bering, Beaufort, and Chukchi Seas, and it didn't take very long for them to accomplish what they had already done in the eastern Arctic. They pared the bowheads down to a precious few. And the whalers might have taken them all, but for the timely invention—and intervention—of lamps that could be lit by natural gas and, later, Tom Edison's electric lightbulb. (How, without baleen, the ladies managed to stay inside their corsets, I am not qualified to say.)

Commercial whaling of other cetacean species, however, continued well into the second half of the 20th century. The gray whale is now extinct in the Atlantic (and probably was extinct there more than a century ago). The blue whale numbers less than five percent of the population it likely mustered before exploitation. The worldwide catch of all whale species peaked in the 1950s and 1960s. At the time, the whaling nations were killing 50,000 a year. Today, apart from such quota-driven aboriginal hunts as the Inupiat's, the only species legitimately taken in international waters is the abundant minke, by Japan, which claims it needs some 400 harvested whales each year for scientific research.

THERE IS A LOT OF ANCESTRAL MEMORY linking people to the sea. And why not? Life started there. We started there. Sure, the branch of life that would lead to humans was a long time out of salt water when it developed a taste for meat and learned to stalk game on the open savannas. But the early ones also stalked easier game in lagoons and salt marshes and tidal pools. And the racial memory of that experience has to be part of the tug-and-pull that keeps drawing us out to the edge of the sea.

"My father saw the ocean as a plate," an old Yupik Eskimo man said to me one day in Chevak, Alaska. Chevak is one of a score of Yupik villages located near the Bering Sea in the Yukon-Kuskokwim Delta, south of Inupiat country but north of the islands of the Aleuts.

The old man's name was Leo Moses. It was lunchtime, and we were sitting at his kitchen table. The tips of his fingers rested lightly at the edge of a large plastic dinner plate. In the center of the plate sat a bowl of fish soup and the dried breast and thigh of a ptarmigan. "The ocean was a plate," Moses said again, "and you always took the first bite from the edge that was closest to you." I watched as his fingers tiptoed across the plate and fastened upon a piece of the bird. "We were told never to finish all the food that was in the plate," he said. "The big one out there, the ocean. We were told not to do that, so there would always be something left for tomorrow." Moses stared at the meat of the bird and the soup of fish, and then

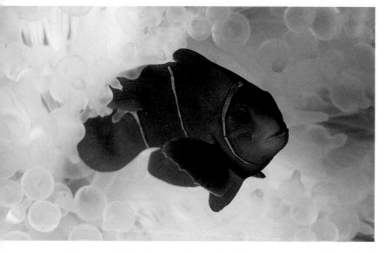

Spine-cheek anemonefish, lurking in its anemone home in Kimbe Bay, Papua New Guinea, will change gender in its lifetime. When this male's female partner dies, it will change from orange to maroon and mate with the next dominant male.
DAVID DOUBILET.

he said, "Excuse me," and ate both. I remember watching him eat and wondering if that apology had been offered only to me, or to the bird and the fish.

I wrote about Leo Moses of Chevak, Alaska, some two decades ago, and I doubt if a single year since has slipped by without some reminder of the wisdom of his message. For now we have been warned with terrible facts from all the frail seas; and the facts from the Bering—accountings of declines in commercial fish stocks and some marine mammals—may turn out to be the most terrible of them all. Excuse me. Have not enough of our fathers seen the ocean as a plate—a plate on which wise men leave something for tomorrow?

About 500 miles north of Chevak, beyond the Bering Strait in Inupiat country, the village of Kivalina sits at the mouth of the char-rich Wulik River, looking out on the shimmering plate of the Chukchi Sea. I was there one June as the sea ice began to rot and drift in floes, with jagged leads winding among them for two or three miles offshore. You could actually hear the melting ice return itself to the glass-smooth sea. It was a sound as from a thousand fountains, dripping—a sound that signaled to the villagers the time for their very best hunters to be out in double-hulled wooden dories, going for *uguruk*, the bearded seal, the fatted pig of the North.

What I remember most vividly of that long-ago day in June, in the blue-green leads offshore from Kivalina, is the hunters with their ice-white parkas and light, bolt-action .25/06 rifles, and then—the outboard stilled, the dory drifting, the floe-fountains dripping—suddenly sighting, 50 or 60 yards out, the round snout of an uguruk pasted like a brown bull's-eye against the ice. What I remember most vividly of all is how the day ended when our boat returned to the gravel beach at Kivalina with a ton and a half of bearded seals, five of them side by side under a tarp. In my mind's eye, I see the hunters attaching ropes through slits in each uguruk's jaw and flippers, the villagers heaving and pulling, the great silver animals sliding reluctantly over the gunwale and up the wide beach to a bench of grass where women are waiting with their sharpened *ulus*.

I remember that evening, above the beach, with the ulus reducing each uguruk into smaller and smaller pieces—the head, the ribs, the blubber, the flippers, even the scooped-up blood. The people had a use for every single part of each animal. Though this happened many years ago, I want to think—and hope—that days and evening like it were still possible to experience in Kivalina last June, and will be possible next June and all the Junes after for as long as the people of Kivalina have the need and inclination to go for uguruk. For if the Kivalinans can think of the ocean as a plate, after the fashion of Leo Moses of Chevak, the take of bearded seals will be sustainable—provided, of course, that American and Russian factory trawlers do not foreclose the subsistence hunt by starving the bearded seal out of existence, scooping its prey from the Chukchi Sea.

At the other end of the western seas, far from the Chukchi in the deeper waters of the Pacific off New Zealand, other commercial fishermen have been scooping up a kind of fish rarely harvested before 1975. Now the orange roughy is on ice at practically every upscale fish market in the United States. On ice and, according to deepwater photographer Emory Kristof, "on the rocks." The succulent roughy inhabits waters a half-mile down. Says Kristof: "We're blowing holes in a food chain we don't yet understand." More than a few marine biologists agree. In fact, some say roughies may live longer than humans, reaching sexual maturity only after 30 years. If that's the case, will the marketplace, even with some harvest restrictions now in place, give this fish the time it will need to replenish its stocks? Don't count on it.

The pioneering bathysphere explorer William Beebe wrote in NATIONAL GEOGRAPHIC in 1931 that the level of deep-sea biological knowledge then available was akin to "the information of a student of African animals, who has trapped a small collection of rats and mice but is still wholly unaware of ... elephants." Now, 70 years later, we are aware of most of the "elephants" of the deep sea, but we still don't know much about them. Emory Kristof is among the latter-day Beebes determined to correct that. A photographer who made the grade at the GEOGRAPHIC by helping to find the *Titanic* and covering other underwater adventures, Kristof allows, "Shipwrecks are fine, but it's the biology down there that fascinates me."

That fascination has taken him—or his equipment, in the form of "rope cams" (photo rigs lowered on lines) and ROVs (submersible, remotely operated vessels)—about as deep as any photographer so far has managed to trip a shutter. With David Doubilet working the shallower water, Kristof has plumbed the depths of Japan's Suruga Bay, and with Bill Curtsinger, he has tested the once radioactive waters of Rongelap Atoll, a hundred miles downwind from Bikini. In 1998 he lowered a rope cam into the mile-deep Kaikoura Canyon, off New Zealand's South Island, to see if he could capture an image of a sperm whale preying on a giant squid. He couldn't, but there's always next time.

Sometimes regarded as a mythical beast, no thanks to Jules Verne, the giant squid over the years has in fact deposited its huge carcass on many a beach. The largest ever recorded washed up on New Zealand sands in 1880. It was measured at a length of 60 feet, and its eyes were each as large as an adult human head. "We know the giant squid is real," says Kristof. "It's down there. It's no Sasquatch. The giant squid says it all about what we still don't know about life in the deep water." ✐

Dwelling on the bottom in warm ocean waters, the saw shark uses eye-appearing nostrils and long nasal barbels to find buried prey, then saws into its intended meal.
DAVID DOUBILET.

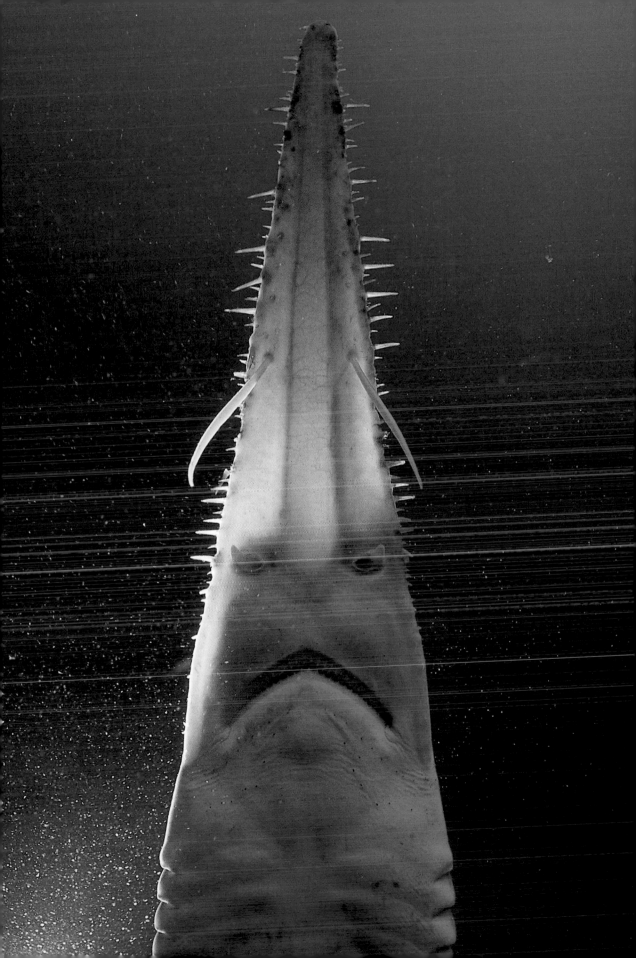

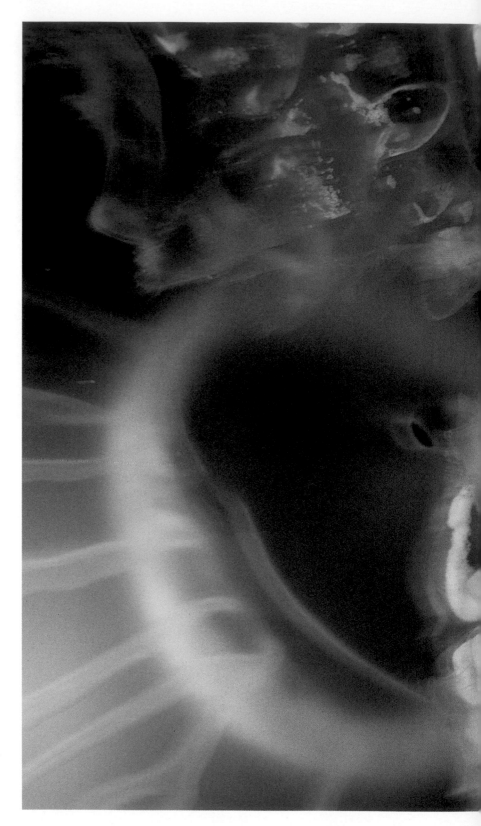

Vying for pastel translucency, a small fish mingles with potentially deadly tentacles of a jellyfish in Tasmanian waters off the southeast coast of Australia.
DAVID DOUBILET.

FOLLOWING PAGES: *Razor-toothed great barracuda, potentially six feet long, leads a pack of crevalle jacks in waters off the Cayman Islands. Both species are carnivorous.*
DAVID DOUBILET.

A tiny goby perches amid the artfully patterned mantle, or flesh, of a giant clam in Indonesia. In shallow water the goby's fused ventral fins serve as suckers, helping it cling to its chosen spot.
DAVID DOUBILET.

Yearling saltwater crocodile grazes the surface of the Jardine River on Australia's northeastern tip. There the olive-brown hunter rules, its smile revealing thick teeth, different from the freshwater croc's needle-thin dentition.
DAVID DOUBILET.

Weedy sea dragon paddles through southern Australian waters, its leaflike appendages providing camouflage. The females of this exotic and fragile-looking species transfer new eggs to males, who carry them under their tails until they hatch.
DAVID DOUBILET.

FOLLOWING PAGES: *Newborn sea horses explore vegetation in the Pacific. An adult male carries developing eggs in a brood pouch in his abdomen, where they grow for several weeks before hatching.*
GEORGE GRALL.

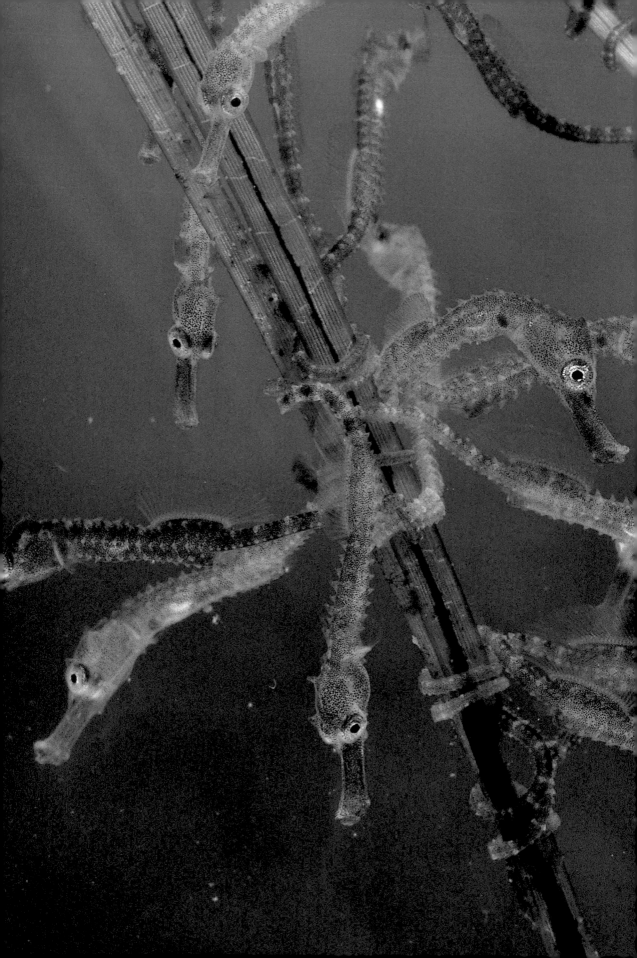

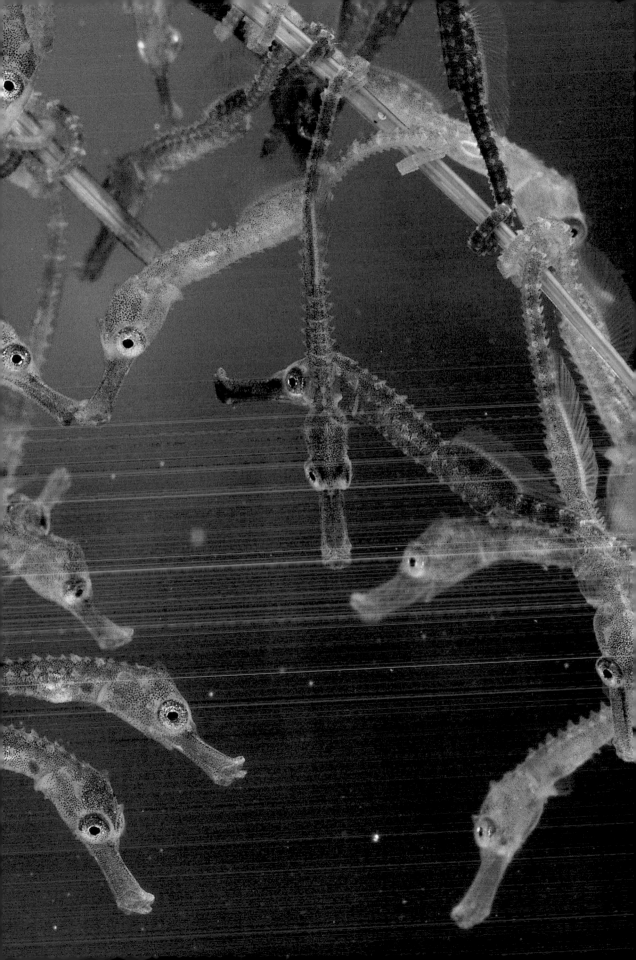

Like charmed snakes, garden eels rise from burrow homes on the Red Sea floor. These eels drill into the sand and secrete a substance to keep holes rigid. Until predators approach, they dance vertically but rarely exit the holes.
DAVID DOUBILET.

FOLLOWING PAGES: *Feather star feeds on passing microorganisms from its seawhip coral home in the Pacific. Suspension feeders, these echinoderms catch floating meals with the sticky appendages on their feathery arms.*
FRED BAVENDAM.

bill curtsinger
& sharks

BILL CURTSINGER IS WILLING TO SUFFER for his art. An accomplished
diver, his special talent is taking a camera underwater. Recently he carried
lenses into the coldest waters in the world—the deeps beneath the thick ice
of Antarctica. There he photographed emperor penguins, Weddell seals, giant
isopods, krill, Antarctic ice fish, and a leopard seal, a creature that can grow
twice as long as a tall human being and is often considered the fiercest of all
bottom-of-the-world predators. To get to the world inhabited by these
creatures, he drilled through 5 to 10 feet of solid ice, then dove down 120 feet
or more in water that never warms above 35°F. (Photographs from the expedition appear in the book, *Life Under Ice*, with text by Mary M. Cerullo.)

Curtsinger hasn't always headed for cold water. In the 1990s, he used his
diving skills in tropical climes, when NATIONAL GEOGRAPHIC posted him to
Bikini Lagoon in the Micronesian Pacific to photograph the sunken warships
of Operation Crossroads, a nuclear test conducted in the Marshall Islands in
1946, the same year that the photographer was born. "Prepared for a waste-
land," he wrote later, "I had found reefs swirling with life. The marine sys-
tems, once bombarded by radiation, had been flushed by time and tides." For
Operation Crossroads, the U.S. Navy positioned 90 target ships in the blue
lagoon. Over the next twelve years, nearly two dozen nuclear test devices

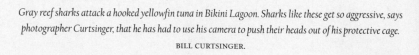

*Gray reef sharks attack a hooked yellowfin tuna in Bikini Lagoon. Sharks like these get so aggressive, says
photographer Curtsinger, that he has had to use his camera to push their heads out of his protective cage.*
BILL CURTSINGER.

Great white sharks swim among a school of herring off the coast of Australia. Twenty feet long, these beasts are sharpsighted yet explore their environment through taste.
BILL CURTSINGER.

Tiger shark swoops up on a black-footed albatross in the Hawaiian Islands National Wildlife Refuge. Skirting the water's surface, albatrosses often fall prey to roaming tiger sharks during fledging season.
BILL CURTSINGER.

were detonated in the area, including a hydrogen bomb a thousand times more powerful than the weapon that nearly vaporized Hiroshima.

After completing the NATIONAL GEOGRAPHIC assignment on the sunken warships, Curtsinger returned to Bikini Lagoon to photograph a different kind of warship: the gray reef shark. "It is a medium-size shark," Curtsinger wrote in his January 1995 cover story, "growing to about six feet. It is also one of the most aggressive. . . . Whenever it feels its space is being threatened, it drops its pectoral fins straight down, raises its snout, arches its back, and starts swimming with an exaggerated weaving and rolling motion. This startling display serves as a warning for an attack."

Curtsinger knew what he was talking about. Two decades earlier, snorkeling alone in a remote Micronesian lagoon in the Caroline Islands, the young photographer was hit by a gray reef shark that came at him "like a rocket." The shark's first pass tore open his left hand and knocked off his face mask; the second raked his shoulder. A friend in a dinghy pulled him to safety.

"I was lucky," Curtsinger wrote. The "defensive animal, I realized later, was not trying to eat me; it was in fact driving me away, possibly seeing me as a potential competitor or predator." Learning from the sharks, the photographer began adopting some defensive behavior of his own. On occasion he slipped into a steel mesh suit inside a plastic "shark scooter." But to get images of sharks in a feeding frenzy, he used a remote-controlled underwater camera, wired to his boat and towing a retractable line baited with a fishing lure called a jet-head. Catching a fish along the outer reef slope was easy. As Curtsinger described it, "once the fish was on, a wave of sharks was never far behind."

In 1979, Friends of the Earth published a handsome volume titled *Wake of the Whale*, with text by Kenneth Brower and photographs by Bill Curtsinger. In it, Brower described his collaborator as a "primate trying to become a seal," a man whose "skin is the thin, Nordic sort intended for use in high temperate latitudes."

Another writer, John McPhee, whose work on the New Jersey Pine Barrens inspired a terrestrial collaboration with Curtsinger two years later, announced much admiration for the photographer's undersea accomplishments. In an illustrated edition of *The Pine Barrens*, McPhee wrote: "A hundred and forty feet below the surface of the Antarctic Ocean, [Curtsinger] once photographed—head-on and close up—a leopard seal. These big creatures have teeth like hatchet blades. They eat penguins as if penguins were salted peanuts, and they will readily attack human beings."

Curtsinger got his salty start in the U.S. Navy. During the Vietnam War, he was a member of the elite Atlantic Fleet Combat Camera Group. He graduated from the Navy Dive and Parachute schools and covered carrier flight operations. Turning to freelance work in 1970, Curtsinger tackled picture stories for a number of magazines, including *Time, Newsweek, Audubon, Smithsonian,* and *Natural History*. For NATIONAL GEOGRAPHIC magazine, his credits list more than 30 articles. Few Curtsinger stories, however, have aroused such fascination as his "Close Encounters with the Gray Reef Shark," published in the January 1995 issue.

After his first trip to Bikini, friends would ask, "Do you glow in the dark, Bill?" The photographer's response: "I tell them that I glow inside, remembering as I do the exquisite times spent diving in the lagoon. My favorite moment came when the largest school of jacks I have ever seen surrounded me, thousands of silver bodies eclipsing the sun." ◄

tropical
forests

ONCE UPON A TIME—A TIME, LET US SAY, within the memory of anyone over 60—tropical forests covered more than a fifth of the Earth's terra firma. They flourished along the Equator and outward to the tropics of Cancer and Capricorn; rolled down the slopes of the Andes, Kenya's Mountains of the Moon, the high ridges of Sumatra and Borneo and New Guinea; spilled across the vast alluvial basins of the Amazon and the Congo; smothered the rolling hills of Southwest India and Southeast Asia; speckled the islands of Melanesia, Polynesia, and the Caribbean. They came in all sorts of configurations and types, from highland cloud forests to lowland rain forests and, here and there for good measure, a dry forest, as on the Mato Grosso Plateau of Brazil and the Deccan Plateau of India. And together these forests sheltered and fed more than half of the Earth's terrestrial species—more kinds of plants and animals than could be found on the rest of the planet's land surface.

The numbers are dazzling still. Worldwide, tropical tree species are believed to total 50,000. On a two-acre site near the Rio Napo in Peru, 300 tree species alone have been counted. A plot only slightly larger, in Ecuador, is said to harbor 50,000 species of insects. The crown canopy of a single tree might sustain 50 species of ants. Within some of the most

The three-toed sloth—easily distinguished from the two-toed variety by a count of its remarkable digits—
moves slowly through the tree limbs in search of fruit and leaves. This one was spotted in Bolivia.
JOEL SARTORE.

diverse habitats, one square mile of rain forest can provide for more than 100 species each of mammals, reptiles, and birds. Imagine what the gods might have counted in 2.5 million square miles of forest, an area ten times the size of Texas. That, more or less, was what the forests of the Amazon basin covered just the day before yesterday, or perhaps it was the day before that. One can be certain of very little in the tropics nowadays. The numbers keep slipping.

There is a saying that the more we risk losing a place, the more we are likely to appreciate it. So it is with these rich and wondrous forests that are disappearing from the middle latitudes of the Earth. And I suspect that, apart from ecologists, no other group of people are so well equipped to appreciate the resource and understand what's at stake than the photographers whose works are represented in this chapter. They have been into these forests on intimate terms, spending days, sometimes weeks at a time, waiting for that one split second when everything is right—the light, the shadow, the direction of the wind, the movement of an animal.

I am thinking of one of those photographers right now: Michael "Nick" Nichols (whose work is profiled starting on page 148). That's him, up there in that tree house blind overlooking a *bai*, or clearing, near the Lokwe River in the Democratic Republic of the Congo. Dawn is breaking. Nichols has trekked through the dark forest from his base camp, getting an early start to avoid spooking the animals that may arrive on the bai when daylight comes. He will be up there for ten hours. Maybe tomorrow he'll spend ten hours in a blind at the foot of the tree house. He will do this for six weeks. He will photograph a bongo, rarely seen in these parts, and forest elephants, and lowland gorillas—scores of gorillas that gather here and turn this clearing into a kind of primate village green. The published photographs will send a message: Don't mess with this. Let it be.

Unlike Nick Nichols, I am a stranger to the tropical forest and its wild creatures. Over the years, assignments afield posted me from Florida's subtropical cypress swamps to the tundra edge of the Arctic Sea, but never into this other place we used to call jungle before that word went out of fashion and, among certain ecological purists, became politically incorrect. On one occasion, however, I did come close to the rain forest. Hitched

Tropical Forests

a ride in an ancient two-engine C-46 fitted for cargo, out of Bogotá in the morning, back in the afternoon from Leticia where, on the River Amazon, the fallen arch of Colombia is jammed between the nose of Brazil and the ear of Peru. For almost half the distance we were over the forests of the Caquetá and Putumayo watersheds.

Along the way, a big anvil-head cumulonimbus, filled with angry lightning, nearly shot us down. I imagined a crash landing in the tree-tops, there to be greeted, should we be so lucky to survive, by a welcoming committee of bird-eating spiders and monkey-eating birds. The experience was denied me. But I saw the forest, saw it sweeping into the east like a great tufted carpet, so dark in the harsh light that it appeared almost black, formidable and uncompromising, hiding answers to questions I would not have the opportunity to ask.

This was in 1978, and Brazil was on a roll to tame the wild Amazon with highways and then use them to resettle the nation's urban poor. The Transamazon Highway was already on its way to Cruzeiro do Sul to rendezvous (though only recently) with Peruvian pavement. But the big road wasn't going to be left out there all by itself. A vast network of highways would soon be under construction, probing the forest from the north and the east. And wherever the bulldozers went, subsistence

The tropics—that great waistband of Earth—once contained forest lands representing more than 25 percent of the global surface. Human population growth and economic development have shrunk tropical rain forests to a fraction of that size in less than a century.

123

farmers, cattle ranchers, gold miners and the sawyers of tall trees were certain to follow. In the western Brazilian state of Rondônia the human head count went from 110,000 to 1,000,000 in just 11 years (1975-1986), while slash-and-burn upped the ante of deforestation from 500 square miles to nearly 7,000.

Beyond the Amazon, tropical forests were receding at a rapid rate. On the plains of the Niger, the Ganges, and the Mekong, millions of people had replaced billions of trees with their villages and pastures and irrigated fields. Java and Sumatra would become so crowded with humanity that the Indonesian government would have to shovel some of the overflow into its territories on the islands of Borneo and New Guinea. Throughout Indonesia, foreign logging companies moved into the lowland forests to harvest the huge, towering, broad-leaved dipterocarps, a family of hardwoods with superior working properties as lumber and much marketplace buzz when sold, misleadingly, as "Philippine mahogany." The lesser stems left standing soon were toppled in the pursuit of slash-and-burn agriculture.

And Africa? Africa's tropical forests—once second only to the Amazon's—are getting to be a mess. Burgeoning human numbers and rapacious European timber interests have rolled back the once-extensive Guinean forests of West Africa, and only a lack of population density and access have temporarily spared Central Africa from a similar fate. Though much of the Congo forest remains intact on the west side of the basin, incessant civil warfare and a consequent pileup of refugees from Rwanda threatens that forest's integrity to the east.

Now, from a historic total of 25 percent of the land worldwide, tropical forest cover has shrunk to a mere 5 percent. Some scientists believe that the loss is running up to 70,000 square miles a year—an area about the size of Cambodia or Washington State. One of the most pessimistic scenarios has the rate of loss peaking in 2010 at about 100,000 square miles a year, after which the numbers would surely plummet, and for a simple reason: There just won't be enough forest left to sustain the decline.

In an excellent work, *Diversity and the Tropical Rain Forest*, John Terborgh of the Center for Tropical Conservation at Duke University writes that there are many primary causes of deforestation around the world. Among them he cites "overt government development policies, often backed up by subsidies and tax breaks," "powerful business interests that regard forests and other natural resources as sources of quick profits," "a lack of alternative energy sources that compels people to cut forests for fuel," and "an insatiable global market for tropical hardwood." Ultimately, he concludes with scalding bluntness, there is the "overwhelming force of ever increasing human numbers, as manifested in the hundreds of millions of would-be farmers who have nowhere to go for land but ever deeper into the forest." And where does that leave the wildlife?

Male red-knobbed hornbill peeks out from the leafy rain forest canopy of its native Indonesian island of Sulawesi.
TIM LAMAN.

It doesn't take an advanced degree in zoology to understand the effect on wildlife of habitat loss. Take away the tropical forest cover and you take away the ability of the birds and insects in the canopies, and most of the fauna on the ground, to feed and reproduce themselves. And as wild forest habitat shrinks into scattered fragments and isolated patches, some of the animals at the top of the food chain suddenly find themselves out of provender. There's a new predator working over the diminished prey base, and it looks—not surprisingly—like a human being.

The name of the game is bush meat, and it is pursued most vigorously in West and Central Africa. Lowland gorillas, chimpanzees, and other primates; duikers and other antelope; elephants, cane rats, lizards, bush pigs—anything that moves is considered potential bush meat. Though laws protect the apes and the elephant, enforcement is rare.

Now, where backcountry subsistence people with antique rifles once took the animals for the family or village pot, commercial hunters armed with automatic weapons stalk the forest trails and logging roads to supply a growing bush meat market in many of Africa's major cities. There is a voracious appetite for bush meat in the logging camps as well. It has been estimated that as many as 500 lowland gorillas are slain for meat every year, just in the Congo. And that's only one small part of a volume that may annually total, in Africa, as much as a million tons of wild meat, the equivalent of four million cattle.

Not all of the bush meat stays on the continent. According to Russell Mittermeier, the primate scientist who heads Conservation International, gorilla is on the menu of a certain restaurant in Western Europe, and 20 suitcases filled with smoked West African meat were confiscated some time ago on arrival in New York City.

Jane Goodall, recently a National Geographic Explorer-in-Residence, told one interviewer that she feared the Africans were eating their future. When the wildlife is gone, she said, many of the indigenous forest people who rely on it for protein may well face starvation. The United Nations Food and Agriculture Organization has voiced similar concerns, adding that declining populations of forest animals could also lead to a long-term change in tropical forest ecology, inasmuch as some plants depend on wildlife for pollination or seed dispersal.

Elsewhere in the tropics, hunting for the pot, protection of livestock, and poaching for bones and other parts fashionable in the aphrodisiac trade have taken a huge toll on the large forest cats. In Indonesia, for example, the tiger was extirpated from Bali and Java years ago. In Laos, the forest is largely intact but it appears that the tiger is essentially gone here, too. Indigenous people have been snaring and eating its principal prey, the barking deer. (In Russia's Far East, obviously outside the reach of the tropics but not outside that of a wondrous feline subspecies, the Siberian tiger is in trouble for a similar reason: Impoverished Russians are hunting and eating elk and boar off the top of the cat's own bill of fare.) A hastily scrawled caption note in the files of photographer Nick Nichols, who spent more than two years among Bengal tigers in India, cuts to the core of the issue: "The main conservation tool [in saving the tiger] should be to preserve prey in adequate densities."

That other large forest cat, the jaguar of Central and South America, likewise faces a troubled future. As humans continue to encroach on its territory, fragmenting habitat and scattering prey, the jaguar is forced to fall back on its opportunistic instincts, and those instincts can lead it to livestock. "Whenever a cow disappears in Latin America," says photographer Steve Winter, "you can bet someone will blame it on the cat. Of course, there aren't enough jaguars left in the wild to account for all the lost cattle. Still, if the owner of a dead cow can track down a live jaguar, that's it. There's going to be one dead cat."

OVER THE LONG HAUL, IF ANY OF THESE charismatic species are to survive the destruction of their habitats, the decimation of their prey, and the antipathy of people "who have nowhere to go for land but ever deeper into the forest," it will likely be in large part because pictures—the work of wildlife photographers—will have helped turn the reluctant human mind to the task of protecting what little remains of the tropical wild. In fact, the mind turning has already begun.

A number of important parks and preserves have been established in the tropics in recent years, only after scientists and photojournalists managed to convince government bureaucrats, with words and pictures, that protection would be in their national interest. In the Congo, for example, the million-acre Nouabalé-Ndoki National Park was established in 1993, largely through the efforts of Michael Fay, a conservationist with the Wildlife Conservation Society and a Conservation Fellow with the National Geographic Society. NATIONAL GEOGRAPHIC's Nick Nichols supported Fay's effort with award-winning photographs of lowland gorillas and forest elephants.

But not all photographers have honed their talents in the tropics working to protect such megafauna as gorillas and forest elephants. Consider, for example, Mark Moffett, who prefers to climb into the tree-tops or crawl upon the ground in search of spiders and beetles and ants. Or Mattias Klum and Frans Lanting, who have photographed their own ample shares of elephants and apes, but who are just as likely to be found scoping the forest for reptiles and birds. Or Tim Laman, the stalker of nocturnal forest gliders.

Moffett is a Harvard-trained ecologist as well as a photojournalist, a man who, in the words of his mentor, the naturalist E. O. Wilson, "thrives in sweat-soaked clothing on torrid afternoons when others fold."

Of the tropical forest, Moffett has written: "Maybe the pulse of the organic world beats stronger for me there than anywhere else." Possibly his own pulse beats stronger there as well.

In his roving assignments on five continents for NATIONAL GEOGRAPHIC and other magazines, Moffett has survived the experiences of accidentally sitting on a fer-de-lance, Peru's (and the Western Hemisphere's) most venomous snake; being driven from a treetop in Colombia by a spectacled bear; crawling through an Aztec tomb guarded by giant spiders; dining on scorpions in China; and living for six months in India on a hundred-dollar traveler's check (a deprivation for which the National Geographic Society takes no responsibility). As for that fer-de-lance, Moffett claims that the secret of his good fortune was to have sat squarely on the snake's head, rather than its tail, thereby immobilizing its fangs. The snake survived.

Passion vine butterfly wings quiver on a floral perch in Costa Rica. Named for one of its food sources, this insect depends on the leaves of the passion vine throughout its larval stage.
DARLYNE A. MURAWSKI.

Tim Laman, another Harvard scientist, has occasionally been to the treetops with Moffett, but more often on his own. In Borneo, Laman photographed forest gliders including flying lemurs, a squirrel the size of a T-shirt, and the paradise tree snake. Why so many different gliding species in Borneo—more than 30—when South American forests sustain none and Africa's harbor but a precious few? Laman believes it is because Borneo's forests are dominated by those huge dipterocarp trees that fruit irregularly at five-year intervals, forcing fruit-eating animals to forage more widely for their food. Evolving the ability to glide, says Laman, solved the tiresome problem of "making the long trip down to the ground and back up again."

FRANS LANTING, A NATIVE OF THE NETHERLANDS, has spent much of his career in the tropics, sometimes with mixed emotions. The tropical forest, he has written, may be a paradise for the naturalist, but it is a photographer's nightmare. Harsh light, deep shadows, lenses fogged by high humidity, equipment smeared with mold: This is a place, he told book editor Leah Bendavid-Val, that is "all blood, sweat, and leeches." One has the sensation, he said, of working in a basement, "rummaging around in semidarkness while the inhabitants are one or two levels up." Still, much of Lanting's finest work has come out of that basement.

Mattias Klum is a native of Uppsala, Sweden, a place even more removed from the tropics than Lanting's home turf. Only 20 years old at the time, Klum got his first taste of the rain forest in a remote part of Sarawak, Borneo. His assignment was to follow a blowgun hunter off the beaten track for three months. He did that, and got hooked. Next stop: Nigeria. Then the Amazon. Then Thailand. Then Borneo again.

Klum is attracted to tropical forests by the challenges they force the photographer to accept. "It's a tough environment," he says, "and it doesn't give you anything for free. But if you can appreciate small wonders while everything about the forest is fighting you, then you will receive the gifts—the great satisfactions—that only this forest can give." Among Klum's recent assignments for NATIONAL GEOGRAPHIC was a photo essay on the king cobra of India and Asia, the largest poisonous snake in the world. Describing its length, the photographer paces off 15 feet. And then, with a smile, he tells you that "some crazy scientist figured out that one king cobra packs enough venom to kill a hundred people." (Fortunately, a cobra can bite only one person at a time.)

Among the many occupational hazards of photographing wildlife in the tropics, snakebite is the least of them. Far more hazardous is tropical disease. Klum contracted cerebral malaria in Nigeria and was sick for half a year. "It almost killed me," he says. Nick Nichols, recording the rigors of Michael Fay's yearlong transect of Central Africa, came down with hepatitis and malaria at the same time. A few years earlier, again in Central Africa, Nichols had to cope with typhoid in addition to those other diseases. "There's a lot of bad stuff out there," he says, "and it can take a lot out of you."

One of the most trouble-plagued assignments in recent years was photographer Joel Sartore's journey into the heart of Madidi National

Park, a Bolivian rain forest so highly charged with aggressive critters that the photographer's field notes soon began to read like a tale from the crypt. On his first day in camp, Sartore noted how his Bolivian host nonchalantly—and at the dinner table—proceeded to dig a living botfly maggot out of her leg. A few days later, the photographer touched a moth and then wiped the sweat from his face. It didn't take too long before he felt the consequences.

In his notebook, he wrote: "I spend the next few hours with my face and hands on fire. Bugs here are toxic . . ." And later: "Juan is up all night, screaming and writhing in pain from parasitic worms in his stomach." Before their visit was over, the Madidi expeditionaries had lost one horse to a jaguar, another to snakebite, and a mule had been badly bloodied— but not downed—by vampire bats.

Months later, back in the States, Sartore was diagnosed with leishmaniasis. In the forest, an insect bite had injected a flesh-eating parasite into his leg. The resultant infection then spread to his lymph system and created a hole in his leg the size of a silver dollar. Surgery and intravenous treatment eventually brought the disease under control.

WOULD SARTORE DO IT ALL OVER AGAIN, knowing the risks? "For a different story?" he says. "Sure I would. We're in a race against time. We're losing species every day. The thing is, I want people to ask themselves, 'What do we choose? Do we choose to save things or do we disregard everything, believing we are above and beyond needing the natural world?' I want to make people think about that."

Ask other wildlife photographers why they are willing to risk their health, if not their lives, in this unfriendly forest environment, and you are likely to hear a similar rationale.

"The key thing is the story," says Mark Moffett. "I want to tell stories that make people excited about little things like spiders. I want to help people look at the unexpected."

And why does Frans Lanting do what he does? He does it "to turn wild creatures into ambassadors for whole ecosystems."

And Nick Nichols? "I feel like these animals need a voice," he says. "Maybe my pictures can help give them one."

Asiatic elephant takes one muddy step in Borneo, where its kind, distinguishable from other Asian elephants, now numbers in the low thousands. According to legend, these elephants' forebears were given by the East India Trading Company to the Sultan of Sulu in 1750. Today, conservation efforts attempt to keep the line alive.
MATTIAS KLUM.

FOLLOWING PAGES: *Langur monkeys congregate at a watering hole in India. With their straight golden hair, langurs populate India's Western Ghats, revered as sacred emanations of the Hindu monkey deity Hanuman.*
MICHAEL NICHOLS.

Caught in the act, a mongoose glances up from devouring its feathered prey. Imported to Hawaii to reduce rodents, mongooses hunt by day and rarely cross paths with nocturnal rats. Now the mongooses have become pests in Hawaii. CHRIS JOHNS.

ABOVE: *Blue morpho butterfly graces a mossy tree trunk in the Bolivian rain forest. Light reflecting off tiny wing scales can make its wings look blue, but brown wing undersides help it disappear from predators.*
JOEL SARTORE.

OPPOSITE: *Appearing alien long before science fiction existed, the long-horned beetle of French Guiana displays intricate head parts including serrated jaw edges. The rain forests where this beetle thrives receive more than one hundred inches of rain a year.*
MARK W. MOFFETT.

FOLLOWING PAGES: *Blue-nose caterpillars rest in formation on a leaf in Costa Rica. Brilliant in color, these fragile larvae sport some 30 spiked barbs, a protective bumper around their outer edges.*
DARLYNE A. MURAWSKI.

Tiger drinks from shallow moving water in India's Bandhavgarh National Park. The elusively nocturnal creature was photographed remotely, using infrared trail monitoring equipment to trip the shutter.
MICHAEL NICHOLS.

OPPOSITE: *Spiderhunter sucks banana flower nectar. This rare bird lives only in Borneo's Kinabalu Montane Alpine Meadows ecoregion.*
MATTIAS KLUM.

ABOVE: *Tarantula sheds its exoskeleton in Peru. This carnivorous eight-legger may live to be twenty years old, likely inhabiting the same burrow through its entire life.*
MARK W. MOFFETT.

Antlered flies battle in
New Guinea. Harmless half-inch
relatives of the Mediterranean
fruit fly, they war over
breeding territories.
MARK W. MOFFETT.

FOLLOWING PAGES: *National
symbol of Guatemala, the
resplendent quetzal takes flight in
its native rain forest. Also called a
kuk, the quetzal was revered by the
ancient Maya and Aztec, who
valued its brilliant emerald
feathers more than gold.*
STEVE WINTER.

michael nichols
& great apes

MICHAEL "NICK" NICHOLS REMEMBERS his first glimpse of Africa's gorilla territory. He was in a small plane above the Rwanda-Zaire border. The pilot jabbed his finger at the forest below and said, "That's where that crazy American woman lives alone with gorillas." Nichols admired that crazy woman—Dian Fossey—and here he was, gazing down on her camp. "It was the most beautiful thing I'd ever seen. Extinct volcanoes everywhere. Lush forest all around." A year later, he trekked into the mountains to do a full-length fundraising feature for her Mountain Gorilla Project. That led to his first book, *Gorilla: Struggle for Survival in the Virungas*, published in 1989. "And that's what set the hook," says Nichols.

"I started with gorillas you could sit next to," he says, "and wound up with gorillas you could only photograph from a distance on a platform." His early work was with mountain gorillas, friendlier than lowland gorillas, he says, "in part because they have not yet been hunted, but also because of Dian Fossey's hard work and sacrifice."

Over the years, his work has taken him to remote corners of the Earth—such as Borneo to photograph orangutans—but for the most part Nichols has focused on the great apes of Central Africa: the mountain and lowland gorillas and the chimpanzees. In the 1990s, for

Tonga, a three-year-old western lowland gorilla, takes one last glance at the camera—and human civilization—before returning home to the wild
MICHAEL NICHOLS.

example, he began photographing the reticent lowland gorillas at National Park in the Republic of the Congo.

In 1992, Nichols discovered a location in the park where he could get close enough to the notoriously shy animals to photograph them. He built a makeshift platform there and began learning the animals' habits. Since then, the platform has been rebuilt and used by researchers to observe and study the elusive creatures. "It's a breakthrough place to study lowland gorillas," says Nichols. "They're afraid of humans because they are hunted throughout their range."

Nichols joined the NATIONAL GEOGRAPHIC staff after a banner year, 1995, in which he was responsible for a quarter of the magazine's cover stories. Then he launched a two-year project to photograph wild tigers in India, an effort that resulted in another book, *The Year of the Tiger.*

In 1998 Nichols shot the photo on pages 150-51 of lowland gorillas snacking and lounging in the Mbeli Bai, a swampy clearing of Nouabalé-Ndoki, where gorillas come to eat and socialize. Taken with a 600mm lens, the remarkable shot required a 45-minute walk to the remote clearing followed by hours of quiet, patient sitting on the platform. "You just sit there all day and hope too many bugs don't bite," says Nichols.

In 1999, Nichols worked in Gabon's Batéké Plateau National Park. His first day there, he photographed a three-year-old western lowland gorilla, orphaned by poachers near birth. Well-meaning locals had taken the infant ape to a beachfront hotel where they had seen other apes on display as tourist attractions. The owner of the hotel, which catered to local oil workers, named the baby Tonga. Gorilla reintroduction efforts funded by the Howletts Foundation influenced the hotel owner, who released Tonga to the national park.

The day Tonga was released, he sat on a tree limb in his "reintroduction site," pensively sucking on a leaf and confronting his natural habitat with some trepidation. "Tonga's reintroduction site was the farthest from humans he had ever been," says Nichols. Despite his initial discomfort, though, the gorilla didn't go back and sleep in his cage that night. After Nichols took the picture, Tonga returned to the trees.

Nichols returned in 2003, finding that Tonga had completely adjusted to his native habitat and his peers. "He's really social and friendly," the photographer reported. "He's got a good chance of becoming the leader. He's developed into a pretty healthy gorilla considering his mom was shot and then he lived in a hotel for three years."

Recently Nichols traveled with conservationist Michael Fay on portions of his Megatransect project—a 15-month, 2,000-mile trek across the African continent. Fay made his way from the Congo's forests to Gabon's seacoast. Subsisting on powdered manioc and canned sardines occasionally delivered by air-drop to the expedition, Fay and team members—few of whom stuck through the entire expedition as he did—gathered data on populations of wild African animals. Accompanying Fay on his remarkable journey, Nichols had the opportunity to capture shots of rarely seen animal life and behavior: an orphaned mandrill, a hippo surfing at Petit-Loango in Gabon, an elephant swimming in the Mambili River, an orphaned gorilla in the Congo's Odzala National Park. The goal was to learn more about the disappearance of animal species in certain areas of Africa, and the reward was a remarkable two-volume book, *Last Place on Earth* (National Geographic Society, 2005). ⟶

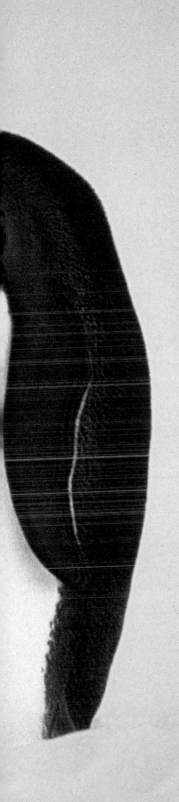

ends of
the earth

AND LAST BUT NOT LEAST WE ARRIVE at the ends of the Earth—the polar antipodes, the Arctic and Antarctic, the ice and rock, the tundra and taiga that bundle up the higher latitudes, the frigid zones, the territories of the barren-ground caribou, the emperor penguin, and the white bear. Raw, lean habitat—the sort most of us would just as soon confront at a great distance, or not at all. So what do we do? We send photographers to the attics of the world to rummage around for us. Let us be grateful for the pictures that come back with them.

What a pity that, without the pictures, popular impressions of these remote places should be attached so firmly to the North and South Poles and to the stories of the adventurers who risked it all to get there first. For myself, I couldn't care less who got there first. Peary, Cook, Scott, Amundsen—it's all the same to me, and to the Poles. Far more fascinating are the skirts of the antipodes, where resident life forms are not limited to lichens and nematodes, and where the stories of exploration say more about practical purpose than personal glory.

The search for a Northwest Passage, for example, had much purpose— a trade shortcut between Europe and the Far East—though not much practicality. The icy passage was finally navigated (by Roald Amundsen)

Emperor penguins regard their year's one and only offspring. They will protect the chick from predatory giant petrels and take turns making the long trek to the sea for fish until the youngster can do so itself.
FRANS LANTING/MINDEN PICTURES.

three centuries after the search began, and one year after construction started on a safer route through the Panama Canal. Still, the effort sprinkled a lot of names across the roof of North America, even as the commercial quest for a critter living under that roof furthered the exploration of northern Canada. Leave it to beaver.

It is astonishing how rapidly the fur trade pulled those early voyageurs across the top of the continent. At a time when Boston Puritans could barely find their way to the Connecticut River, the Frenchman Sieur des Groseilliers was trafficking with Cree on the shore of Hudson Bay. Virginians had hardly scuffed their boots at Snickers Gap in the Blue Ridge when the Englishman Henry Kelsey stood on the barren ground of Saskatchewan, counting musk ox and grizzly bears. And even before Daniel Boone blazed a trail 200 miles from Carolina to Kentucky, Samuel Hearne, in the service of his majesty's Hudson's Bay Company, trekked overland from that bay and up the Coppermine River to the Arctic Ocean, 2,700 miles from Montreal. The legacy of all these subarctic walkabouts, of course, was an ever expanding fur trade that finally crashed in the 1840s, partly because the American beaver was getting hard to find, but mostly because the mad hatters of Europe had discovered an Asian caterpillar, *Bombyx mori*, that spun a lustrous fiber called silk. Thus did the silken hat replace the beaver's felt—and just in the nick of time for the beaver.

At the other end of the Earth, among the outlier islands off Antarctica, seals of one species or another had been luring sailors from half a dozen northern nations to the edge of the ice. They sailed in the wake of the English mariner James Cook, who, in 1773, had missed out on being the first European to sight the last continent. In his disappointment, the captain foolishly predicted that "no man will ever venture farther than I have done." And yet, Cook went on, should someone succeed, "I shall not envy him the honour of discovery, but I will be bold to say that the world will not be benefited by it." But the world was benefited, at least for a while: Sealskin coats soon became as popular as beaver hats. Inevitably, however, the trade in skins had to collapse from the weight of its own excess—just in the nick of time for the southern fur seal.

Now even tourists can get to Antarctica, though the folks who spend real time there tend to be scientists, construction workers (building facilities for the scientists), and journalists toting notebooks and cameras (to record what sort of science the scientists are brewing). Kim Heacox is such a journalist: a photographer, in fact, as noted for his words as for his pictures. In 1997, the National Geographic Society Books Division assigned Heacox, an Alaskan, to see how the ice and the animals were holding up at the planet's other end. This is how he introduced his account in *Antarctica: The Last Continent*:

The Arctic and Antarctic regions may be considered "the ends of the Earth," but they hold within their icy crystalline realms nearly three-quarters of the globe's fresh water. Global climate change is making a measurable impact on these habitats and the wildlife in them.

> If any place on this precious Earth belongs to everyone and to no one, it is Antarctica, the white continent. We invert the world to examine it and become inverted ourselves. We converge upon it with every line of longitude, encircle it with our high latitudes, test it with our science, and find—in turn—that it tests us. It tests not just our science but also our conscience, our poetry, our art, our arrogance, and our assumptions. It contrasts our finest technologies against a single epiphany, a simple adaptation: a seal in the sea or a mother penguin and chick on the ice who call amid a cacophony of 10,000 other penguins, their voices all the same to us, and find each other. We bow before the magic and awaken to a new beginning. It is somehow always morning here.

Nine-tenths of the world's ice—and within its crystals, seven-tenths of the planet's fresh water—is locked into Antarctica. In some places, vertical ice runs three miles deep. During the austral winter, the continent is surrounded by floating sea ice, millions of square miles of it. Winds kick up at a hundred miles an hour. Fahrenheit temperatures plummet to 60 below. It is a puzzlement how anything can live here—and not much does, far from the seaward edge of the ice. But along that edge, and out across the floes toward such islands as the South Shetlands, the region brims with wildlife—skuas and petrels and albatross, seven species of penguin and six of seal, and all of them hitched one way or another to a food chain built on a billion tons of krill.

Of all the Antarctic penguins (indeed, of all the 17 species throughout the world), the cock of the walk is the emperor, a four-foot, 70-pound bird that dives to astonishing depths to feed on fish and squid in frigid waters. On pages 176-77, Frans Lanting has brought us a picture of a crèche of emperor chicks awaiting the return of their hunting parents, and on page 154 he has given us a glimpse of parent affection, emperor penguin-style.

AMONG THE GEOGRAPHIC'S REGULAR CONTRIBUTORS in recent years, none has spent so many mornings at the ends of the Earth as Maria Stenzel, whose photo assignments have taken her four times to the Antarctic, including South Georgia Island; once to Siberia to travel with nomadic Nenets reindeer herders; once to northern Canada to paddle and portage the historic spoor of the fur trader David Thompson; and once to the village of North Pole, Alaska, which is nowhere near the true North Pole—and which she is loath to count, since the gig was short and involved neither landscape nor wildlife. Fen Montaigne, the writer who worked with her in Siberia, correctly describes Maria Stenzel as "a photographer with a penchant for high latitudes."

On her most recent assignment in the Antarctic, Maria Stenzel rediscovered a proclivity for something else—photographing wild animals in their natural habitats. Her previous sorties to the antipodean underworld had demanded focus of an altogether different sort—historical (for a recount of the Shackleton expedition), geological (for a study of the continent's dry, iceless valleys, where the top predator happens to be a microscopic, bacteria-eating nematode), and glacio-logical (for a study of ice). But now she found herself on the loose among emperors, skuas, and Weddell seals. "I loved it," she said on her return to the States after almost four months afield. "It's such a different world at the edge of the ice: not land, but not ocean either. That's where the life is, and there's so much of it."

From the photographer's caption notes, I see that Stenzel spent some time observing Adélie penguins, while noting the bird's unsolicited role as an indicator species—a kind of Antarctic canary-in-the-mine-shaft of global warming.

"At a population of 2.5 million pairs," Stenzel writes, "the Adélie is the most abundant and studied bird of the Antarc-tic." But studies now indicate that its abundance is in jeopardy along the coast of the Antarctic Peninsula, that 1,200-mile-long finger of rock reaching north toward the tip of South America. "Since the 1940s, the average year-round temperature on the peninsula has increased about 4 degrees Fahrenheit," she writes. "This is the greatest warming anywhere on Earth . . . ten times the global average." In the vicinity of the United States base at Palmer Station, Adélies "have dropped from 15,000 breeding pairs to 7,700 breeding pairs in the past 25 years." One biologist told Stenzel that as temperatures warm in the winter, sea ice disintegrates, reducing the algae that live under the ice and fatten the krill that feed the Adélies. The Larsen Ice Shelf in the Weddell Sea, for example, has shrunk by almost one thousand square miles in the past two years.

Peary caribou stride across frozen earth on Ellesmere Island, Canada, part of an Arctic archipelago representing the only habitat that supports this ungulate subspecies.
JIM BRANDENBURG.

THERE IS SOME EVIDENCE THAT CLIMATE CHANGE may be affecting wildlife in the Arctic as well. As seasonal warming pushes the edge of the ice farther from shore, some polar bears could eventually find their primary prey base—seals—unreachable. Photographer Flip Nicklin reports that so far the Alaskan bruins seem to be weathering the warmth. "But over time," he adds, "any kind of climatic change would alter the system for many species, and that could be bad news for the bears." (Read more about Nicklin and his portraits of the great white bear, beginning on page 180.)

Fortunately, the range of this powerful predator is circumpolar, embracing northern Canada, Alaska, Russia, Norway, and much of coastal Greenland, and its total population is said to fall somewhere between 25,000 and 40,000, a number deemed sufficient to keep the animal off any endangered list. Hunting the bear is now largely under control thanks to an international agreement. Norway has banned the hunt on its territory altogether, while Denmark (for Greenland), Canada, and the United States (for Alaska) permit Inuit subsistence hunters to take bears by quota—amounting to an international total of up to about 750 a year.

And of course, given the rare opportunity, a hungry polar bear is fully capable of taking human prey for its own subsistence. Legends from the North Country frequently feature anecdotes of bear attacks on the open ice. And more often than not, the bruin is credited with a supernatural cunning—as in the following tale:

> Once there was a man who refused to believe that some bears are smarter than people. Walking alone through the icy hills one day, he happened to see a great white bear watching him. The man carried a long staff with a tip like the point of a spear, and he bravely shook the staff to show the bear that he was armed. The bear took note of this and allowed the man to pass unharmed. Then, some distance down the path, the man with the staff met another wayfarer heading in the opposite direction. The second man carried no staff or spear, so the first man warned him of the bear ahead and handed him his own staff for protection. Whereupon the second man placed the staff over his shoulder and went on his way, and by and by he too passed without incident under the eyes of the watchful bear. But the bear remembered something—he had seen that same staff once before, going the other way. So the bear got up and tracked the first man down. And ate him.

"I suspect it's a love-hate thing," says Richard Olsenius of his feeling for the Far North. Having spent much of his life in the temperate woodlands of the Upper Great Lakes region, Olsenius felt a sharp disconnect when NATIONAL GEOGRAPHIC magazine first posted him into the stunted spruce latitudes north of 60 degrees. (The assignment that yielded one of the photographs in this book, on pages 178-79—an aerial of caribou on the hoof in Labrador—was his third to the boreal region.) "What struck me that first time was the flattening out of the trees and the country, and the cold and the wind that wanted to flatten you," he recalls. "I looked at the texture of that landscape and thought—wow, is that it, all the way to the North Pole?"

Other photographers whose careers include a longer span of experience in the North can afford to be blasé. Flip Nicklin, for one, has probably spent no fewer than 300 nights of his chilling career perched at the edge of the Arctic ice, waiting for seals or bears or bowhead whales. "It's not a comfortable place, the Arctic," Nicklin reported while on a strikingly different sort of assignment, whale watching in Hawaii. "It's the kind of place where your hands ache."

A quarter century ago I got my first assignment to the edge of the North, though not very far north: Fairbanks, Alaska. It was February, short days and long nights. The saloons and the sidewalks downtown were crowded with prostitutes in leather miniskirts and construction workers in pointy-toe boots. Most of both types had come into the country from "Outside"—Outside being the Alaskan way of dealing with the lower 48. It was boom time. A huge cache of petroleum had been discovered under the North Slope, and those pointy-toe boots were building a pipeline to bring it to market. In a year or so, someone up at Prudhoe Bay was going to open a valve and let 'er gush. But I wasn't in Alaska to snoop around the pipeline. I was there to find out why the state of Alaska was determined to reverse the decline of its moose and caribou populations by waging an aerial war on the wolf.

According to the state, there were too many wolves. Fewer wolves would mean more caribou. To those who argued that fewer hunters— of the kind known elsewhere as "sports hunters"—would mean more caribou, the state was obliged to explain: Yes, but the wolves don't buy hunting licenses. (Neither did some of the pointy-toes in the construction camps.)

Over the years, I have been back to northern Alaska half a dozen times, usually under better light than February's, and each trip placed me a bit farther beyond the Arctic Circle until at last I ran out of traveling room at the edge of the Beaufort Sea. And almost always in the backgrounds of the stories I was sent to cover loomed the same iconic wild species—the barren-ground caribou, *Rangifer tarandus* to the taxonomist, *tuttu* to the Inupiat Eskimo, *vadzaih* to the Athabascan

Gwi'chin, whose tribal name itself means "people of the caribou." After a while it struck me, and it strikes me still, that caribou are the glue that has held this Far North country and its native peoples together for thousands of years.

On Alaska's Kobuk River, within the migratory range of the Western Arctic caribou herd, is a place that is known as Onion

Atlantic puffins perch precariously on a lichen-encrusted rock in Norway. Gawky on land, in the water they plunge, torpedo-like, propelled by muscular wings and wide webbed feet toward their prey. FRANS LANTING.

Portage. It was here some years ago that J. L. Giddings, a scientist from Brown University in Providence, Rhode Island, uncovered the middens and log-igloo pits of an ancient race of caribou hunters—ancestors, perhaps, of the Inupiat people who live today along the Kobuk in such villages as Kiana and Shungnak. Giddings's discoveries here date from a time before any pyramid ever cast a shadow over Egypt. The portage, he deduced, was where, in the fall, great cohorts of caribou surged out of the Baird Mountains to cross the river into their winter range. And the ancient Kobuk people would be waiting for them, with fences of sticks to funnel the animals into the water, where they might be slaughtered with spears from bark canoes.

"I could visualize," Giddings wrote in his classic account, *Ancient Men of the Arctic*, "the plunging of spears and knives into the demoralized swimmers, the spurts of blood mingling with the rush of the blue Kobuk water, and the dead and dying animals, held aloft by their buoyant coats, drifting with the current to the gravel beach at Onion Portage."

The portage today lies within the boundary of Kobuk Valley National Park. It is both lawful and appropriate that Inupiat subsistence hunters from the villages still come here—but now by motorboat, and with rifles, not spears—to await the migrating herd. When I went to the portage two decades ago to camp in the cabin that Giddings had built and to contemplate the significance of his discovery, both the hunters and the soon-to-be hunted were on their way. From the air, flying in, I had seen a few caribou, the lead bulls, out on the tundra plain below Jade Mountain. Their breasts showed white against the purple heath, and the arcing curves of their antlers all pointed in the same direction—toward the river at Onion Portage.

Another great caribou population ranges across the northeastern side of Alaska. It is called the Porcupine herd, named for the river that waters much of the herd's winter range. From the Gwi'chin community of Arctic Village, just off the underside of the Arctic National Wildlife Refuge, I traveled one day with some caribou hunters, poking up the East Fork Chandalar toward the Brooks Range. But in 30 miles, up and back, we saw only a few tracks in the riverside sand—and not a single caribou. It was still too early in the fall for the bulk of the herd to be out of the mountains. Maybe next week.

From the bend in that river where the hunters decided to turn their aluminum punts toward home, we could look northeast past the icy cone of Nichenthraw Mountain to the top of the Brooks Range. Beyond it, the Gwi'chin knew, steep valleys carried the North Slope's rivers—the Jago, Okpilak, Hulahula, Sadlerochit—down and out across the tundra of the coastal plain to the Beaufort Sea, down to the calving grounds of the Porcupine herd, down to where the success or failure of each caribou birthing season would likely determine the future success or failure of Gwi'chin hunters on the East Fork Chandalar.

The coastal plain of Alaska's North Slope is about as close as I've ever been to either end of the Earth, thanks (once again) to the caribou. Well, matter of fact, not just the caribou—thanks to the oil, a reservoir of petroleum believed by some geologists to be the most productive untapped onshore field in the United States, almost all of it underlying the calving grounds of the Porcupine caribou herd on the coast of the Arctic National Wildlife Refuge.

There is an issue here. Can the oil be extracted without harm to the refuge's wildlife? Can pipelines and drill pads—however less intrusive than those built 25 years ago at Prudhoe Bay—be compatible with calving caribou? Or would the industrial infrastructure nudge the herd back into the Brooks Range foothills, where increased predation by eagles and wolves and grizzly bears might cut the rate of calf survival by half?

We have heard these questions debated in the United States almost every time there is a changing of the guard on Capitol Hill or at the White House. To drill, or not to drill. The proponents of development offer strong arguments that there is no problem here: You can have your cake and eat it too; the caribou and the polar bears and the musk oxen, the Inupiat and the Gwi'chin won't even blink. The opponents of development say phooey. Oil and infrastructure; wildlife and wilderness: Take your pick.

Me, I pick wilderness. I pick it remembering my time beside the Jago River, on the coastal plain halfway between the mountains and the sea. The herd was running late that spring, snow still piled in the passes. But we saw a couple of caribou, scarfing up cotton grass across the river, and a grizzly downstream, sniffing the air. I had never felt such distance as I did in the long, open, aching silence of that place, or such solitude. In the sun-bright mornings, looking north, I could see a thin white line dancing above the far horizon—a mirage, a sunlit reflection off the ice of the Beaufort Sea. The Inupiat have a word for this kind of optical illusion. They call it *innipkak*.

And it occurred to me then, as it does now, that the greatness of the United States of America may float like a bit of innipkak, too. For how can a nation be truly great if, in order to pursue the fleeting equivalent of a few months' supply of oil, it is willing to risk the biological integrity of an ecosystem unlike any other in the world? If the planet's richest country will not protect its own wildlife resources, what can we possibly expect for the future of wildlife in the cash-poor, biologically rich ends and corners of the Earth?

Steller sea lion seems to kiss her pup as they lounge together on a rock near the Bering Sea. In western Alaska especially, this species declined precipitously in the 1980s. Rules and research are now in place to save the Steller sea lion. JOEL SARTORE.

Arctic wolf enjoys a spot of
sunlight on a snowy cliff on
Ellesmere Island, a realm of
Canada so nearly devoid of human
life that this creature seems
undisturbed by the photographer.
JIM BRANDENBURG.

FOLLOWING PAGES: *Moose
wanders the Alaskan forest mists,
raising head and voice against
an expansive background that
signifies the resources required for
its sustenance.*
MICHIO HOSHINO/
MINDEN PICTURES.

OPPOSITE: *Atlantic puffin grips a feast of sand lances on Scotland's Outer Hebrides. Unlike the crow in the fable, this puffin need not open his neon beak to talk but can convey messages to fellow birds through gesture and gait instead.*
FRANS LANTING.

ABOVE: *Their numbers now slowly on the rise, the Eurasian crane was rare in the British Isles even during Elizabeth I's reign. At a banquet in 1577, one hundred other wild birds were served, but only one precious crane.*
STURE TRANEVING.

FOLLOWING PAGES: *Underwater glimpse of an Arctic polar bear, captured by remotely controlled camera in the frigid waters of Canada's Northwest Territories.*
FLIP NICKLIN.

Walrus bulls populate an Alaskan island shorefront. This walrus bachelor party goes on for most of the summer; meanwhile, the cows and calves stay home in the pack ice of the Chukchi Sea.
JOEL SARTORE.

Fledgling Emperor penguins crowd together in Antarctica. The one parent among them seeks its young, distinguishing it from all others by its unique call.
FRANS LANTING/
MINDEN PICTURES.

FOLLOWING PAGES:
Barren-ground caribou gallop across the frozen Labrador tundra. Canada's most abundant caribou species, they migrate long distances in massive herds. Five herds are so long-lived and familiar that they have been given names.
RICHARD OLSENIUS.

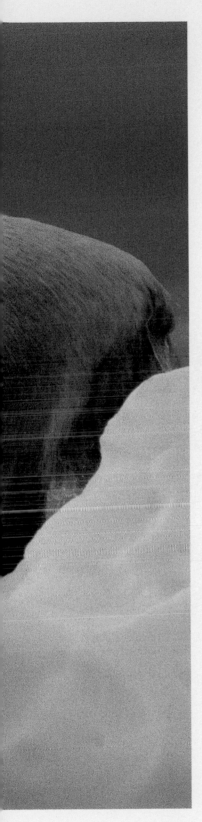

flip nicklin
& polar bears

POLAR BEARS, EARTH'S LARGEST LAND CARNIVORES, use the Arctic Sea ice as a fishing platform for catching seals. And that ice appears to be melting at the rate of about 9 percent a decade. As the time between fall freeze and spring thaw in the Northern Hemisphere grows shorter (nine days earlier and ten days later, respectively, than 150 years ago), the bears are gradually forced farther north. In fact, according to some scientists, global warming over the next 50 years may lead to the extinction of some species—including the polar bears.

That's why the photography of Charles "Flip" Nicklin is so valuable. He documents these soft, fluffy-looking, yet fearsome creatures. The polar bear—great Nanook of the North, *Ursus maritimus*, the white bear in the land of the white ice, the creature that (according to Eskimo legend) casts no shadow—is not the world's largest bear. That title belongs to the brown bear. But since brown bears are omnivores, the polar bear is the largest land-dwelling meat-eater. And it does some of its best hunting underwater.

Flip Nicklin has spent a good part of his life on and under the water with wildlife. "Whale Man," as he has been dubbed, has swum with the best of them—sharks, seals, squid, sea turtles. But not sea bears.

Polar bear shakes off water after a chilly July swim in Canada's Wager Bay, Northwest Territories. Since native Inuit built a lodge catering to ecotourists, this bear has learned not to fear humans.
FLIP NICKLIN.

Young males tussle on the shore of Hudson Bay in Manitoba, Canada. Polar bear play-fights rarely result in injury but do develop physical skills and strength, readying youth for Arctic challenges.
FLIP NICKLIN.

Swim with a polar bear? "Not a very good idea," says Nicklin. Adult males can weigh in at 1,000 pounds, alpha males at 1,500. Of one tranquilized bear in the Queen Elizabeths, NATIONAL GEOGRAPHIC Senior Writer John L. Eliot wrote: "I kept staring at the head, a massive triangular wedge that looked as if it could smash through solid rock." Ringed seals are the bear's most common prey. Although it could eat one of these seals in a sitting, given its prodigious stomach capacity, the bear will generally consume only the skin and the blubber.

Polar bears have a highly developed sense of smell that allows them to sniff out their favorite snack: seal pups cached by their mothers in snowdrift hollows. In summertime, with the ice breaking up, some bears will swim underwater to surprise seals from beneath at their breathing holes. And in the fall, males often pair up for bloodless play-fights—a time, according to one researcher, for the boys "to be social."

To get the remarkable photo of a swimming bear on pages 172-73, Nicklin enlisted the help of Joe Stancampiano, a National Geographic Society photo equipment specialist. Stancampiano rigged up an underwater camera housing and lowered it on a four-foot pole from a small boat. The rest of it was up to the bear—and to Nicklin, who tripped the shutter remotely. (In 1994, in another Society photo collection, Leah Bendavid-Val, the project editor for this volume, wrote: "Stancampiano does not usually go out with Flip Nicklin. Stancampiano hates the cold, and Nicklin likes to work in the polar regions." So how come, Joe? "Aw," he says, "it wasn't all that cold, out of the water.")

For other pictures displayed here, Nicklin traveled to Canada's Hudson Bay and the Queen Elizabeth Islands with Stancampiano and Eliot. Churchill, Manitoba, on the bay, advertises itself as "the Polar Bear Capital of the World," and Wapusk National Park, nearby, is said to have the planet's largest concentration of polar bear denning sites.

It is said of Flip Nicklin that he was born with both diving and photography in his blood. His father, Chuck, an underwater cinematographer, got him scuba diving off the California coast at an early age. In 1976, Flip signed on as a deck hand and diving assistant for a three-month shoot with NATIONAL GEOGRAPHIC photographers Bates Littlehales and Jonathan Blair. With the help of those mentors, he landed his first

assignment for the GEOGRAPHIC: photographing humpback whales. "I love digging into long-term stuff," says Nicklin, who even today continues his study of humpbacks and other whales.

In 1998, Flip Nicklin joined David Doubilet to help celebrate "Blue Refuges," the United States' 12 marine sanctuaries. And the year before that, he slipped into Arctic waters in Norway's Svalbard archipelago to photograph bearded seals, a sometime prey of the polar bear, and came out of it with a close-up, face-to-face cover photo. After snapping the picture underwater, Nicklin surfaced for air. So did the seal.

"All of a sudden I felt a pair of flippers on my forearms, and there she was, nuzzling the top of my head," Nicklin recalled later in a piece for the magazine's "On Assignment" page. In all his years of photographing in the Arctic, he continued, he had gotten only a couple of shots of bearded seals—"and those were from a distance," he wrote. "It was great to get that close at last." While he may never have felt a polar bear nuzzling the top of his head, Flip Nicklin did move close in on these ice-white ursines—and brought the photographs back to prove it. ➤

Polar bear naps to conserve energy in Manitoba, Canada. In the fiercest weather, polar bears make shallow snow bank burrows and doze, paws over their heat-radiating muzzles, to stay warm.
FLIP NICKLIN.

further reading

Annie Griffiths Belt, *Last Stand: America's Virgin Lands*
(National Geographic Society, 2002)

Jim Brandenburg, *Animals of Africa*
(Hugh Lauter Levin Associates, 1997)

Jim Brandenburg, *Brother Wolf: A Forgotten Promise*
(North Word Press, 1993)

Jim Brandenburg, *White Wolf: Living with an Arctic Legend*
(North Word Press, 1988)

Kenneth Brower, *Realms of the Sea*
(National Geographic Society, 1991)

Kenneth Brower and Bill Curtsinger, *Wake of the Whale*
(Friends of the Earth, 1979)

Patricia Caulfield, *Everglades* (Sierra Club, 1970)

Mary M. Cerullo, photography by Bill Curtsinger,
Life Under Ice (Tilbury House, 2003)

Douglas Chadwick and Joel Sartore,
The Company We Keep: America's Endangered Species
(National Geographic Society, 1996)

Bill Curtsinger, *Extreme Nature: Images from
the World's Edge* (White Star, 2005)

David Doubilet, *Fish Face: Portraits* (Phaidon, 2003)

David Doubilet, *Great Barrier Reef*
(National Geographic Society, 2002)

David Doubilet, *Water Light Time*
(Phaidon, 1999)

David Doubilet and Andrew Ghisotti, *The Red Sea*
(Smithmark, 1994)

David Doubilet, *Pacific: An Undersea Journey*
(Bullfinch Press/Little, Brown, 1992)

David Doubilet, *Light in the Sea* (Thomasson-Grant, 1989)

Sylvia Earle and Wolcott Henry,
Wild Ocean: America's Parks Under the Sea
(National Geographic Society, 1999)

Michael Fay, *Megatransect: Mike Fay's Journals*
(National Geographic Society, 2005)

Ron Fisher, photography by James P. Blair,
*Our Threatened Inheritance: National Treasures of the
United States* (National Geographic Society, 1984)

Ron Fisher, photography by Sam Abell and
David Doubilet, *Wild Shores of Australia*
(National Geographic Society, 1998)

James Gorman, photography by Frans Lanting,
The Total Penguin (Prentice Hall, 1990)

Noel Grove, photography by Frans Lanting,
Galen Rowell, and David Doubilet,
Living Planet: Preserving Edens of the Earth
(Crown Publishers, 1999)

Bobby Haas, photography by Kuki Gallmann,
Through the Eyes of the Gods: An Aerial Vision of Africa,
(National Geographical Society, 2005)

Kim Heacox, *Visions of a Wild America:
Pioneers of Preservation*
(National Geographic Society, 1996)

Kim Heacox, *Alaska's Inside Passage*
(Graphic Arts Center Publishing, 1997)

Kim Heacox, *Antarctica: The Last Continent* (National
Geographic Society, 1998)

Chris Johns, *Valley of Life: Africa's Great Rift*
(Thomasson-Grant, 1991)

Chris Johns, text by Peter Godwin, *Wild at Heart:
Man and Beast in Southern Africa*
(National Geographic Society, 2002)

Dereck Joubert, photography by Beverly Joubert,
Hunting with the Moon: The Lions of Savuti
(National Geographic Society, 1997)

Dereck Joubert and Beverly Joubert, *The Africa Diaries:
An Illustrated Memoir of Life in the Bush* (Adventure
Press/National Geographic Society, 2001)

Frans Lanting, *Jungles* (Taschen, 2000)

Frans Lanting, *Penguin* (Taschen, 1999)

Frans Lanting, *Eye to Eye: Intimate Encounters with the Animal World* (Taschen, 1997)

Frans Lanting, *Okavango: Africa's Last Eden* (Chronicle Books, 1993)

Frans Lanting, *Madagascar: A World Out of Time* (Aperture, 1990)

Light on the Earth: Two Decades of Winning Images (Wildlife Photographer of the Year) (BBC Books, 2005)

David Littschwager and Susan Middleton, *Archipelago: Portraits of Life in the World's Most Remote Island Sanctuary* (National Geographic Society, 2005)

John G. Mitchell, *The Man Who Would Dam the Amazon, and Other Accounts from Afield* (University of Nebraska Press, 1990)

John McPhee, photography by Bill Curtsinger, *The Pine Barrens* (Farrar, Strauss & Giroux, 1981)

Mark W. Moffett, *The High Frontier : Exploring the Tropical Rainforest Canopy* (Harvard University Press, 1993)

National Geographic's Last Wild Places (National Geographic Society, 1996)

Michael Nichols, *Last Place on Earth* (National Geographic Society, 2005)

Michael Nichols, essay by Jane Goodall, *Brutal Kinship* (Aperture, 1999)

Michael Nichols and Geoffrey C. Ward, *The Year of the Tiger* (National Geographic Society, 1998)

Michael Nichols, *The Great Apes: Between Two Worlds* (National Geographic Society, 1993)

Michael Nichols, *Gorilla: Struggle for Survival in the Virungas* (Aperture, 1989)

Flip Nicklin and James Darling, *With the Whales* (North Word Press, 1990)

Laura Riley and William Riley, *Nature's Strongholds: The World's Great Wildlife Reserves* (Princeton University Press, 2005)

Joel Sartore, *Nebraska, Under a Big Red Sky* (Nebraska Book Co., 1999)

George Shiras III, *Hunting Wild Life with Camera and Flashlight: A Record of 65 Years' Visits to the Woods and Waters of North America* (National Geographic Society, 1936)

Roff Martin Smith, photography by Sam Abell, *Australia: Journey Through a Timeless Land* (National Geographic Society, 1999)

Page Stegner, photography by Frans Lanting, *Islands of the West: From Baja to Vancouver* (Sierra Club Books, 1985)

Whales, Dolphins, and Porpoises (National Geographic Society, 1995)

Wildlife: The World's Top Photographers and the Stories Behind Their Greatest Images (Rotovision, 1994)

Wildlife Photographer of the Year, Portfolio 15 (BBC Books, 2005)

World Wildlife Fund, with photography by Frans Lanting, Galen Rowell, and David Doubilet, *Living Planet: Preserving Edens of the Earth* (Crown, 1999)

about the author

JOHN G. MITCHELL is a longtime contributor to NATIONAL GEOGRAPHIC magazine. Only recently having left his post as a senior editor, he continues to write extensively for the magazine. Mitchell has held a variety of other positions in the course of his long and distinguished career. He has served, among other things, as science editor at *Newsweek* and as editor-in-chief of Sierra Club Books. He is the author of five other books, including *The Man Who Would Dam the Amazon and Other Accounts from the Field*. He has contributed scores of articles to such publications as *American Heritage*, *Audubon*, *Smithsonian*, and *Wilderness* magazines. In his most recent contribution to NATIONAL GEOGRAPHIC, he teamed up with photographer Joel Sartore for a July 2005 feature on the environmental, economic, and social effects of drilling for natural gas in the American West.

index

Boldface indicates illustrations.

The Wildlife Photographs
John G. Mitchell

Published by the National Geographic Society
John M. Fahey, Jr., *President and Chief Executive Officer*
Gilbert M. Grosvenor, *Chairman of the Board*
Nina D. Hoffman, *Executive Vice President;*
 President, Books and Educational Publishing

Prepared by the Book Division
Kevin Mulroy, *Senior Vice President and Publisher*
Kristin Hanneman, *Illustrations Director*
Marianne R. Koszorus, *Design Director*
Rebecca Hinds, *Managing Editor*
Barbara Brownell Grogan, *Executive Editor*

Staff for this Book
Leah Bendavid-Val, *Editor*
Vickie Donovan, *Illustrations Editor*
Carol Norton, *Art Director*
Suzanne K. Poole, *Researcher*
Carl Mehler, *Director of Maps*
Alicia M. Moyer, *Editorial Assistant*
R. Gary Colbert, *Production Director*
Lewis Bassford, *Production Project Manager*
Meredith Wilcox, *Illustrations Assistant*

Staff for this Edition
Susan Tyler Hitchcock, *Project and Text Editor*
Michael Heffner, *Illustrations Editor*
Linda McKnight, *Art Director*
Susan Straight, *Researcher*
Richard S. Wain, *Production Project Manager*
Teresa Neva Tate, *Illustrations Specialist*
Margo Browning, *Contributing Editor*
Suzanne Zito Slezak, *Proofreader*
Margie Towery, *Indexer*

Manufacturing and Quality Control
Christopher A. Liedel, *Chief Financial Officer*
Phillip L. Schlosser, *Managing Director*
John T. Dunn, *Technical Director*
Vincent P. Ryan, *Manager*
Maryclare Tracy, *Manager*

Founded in 1888, the National Geographic Society is one of the largest nonprofit scientific and educational organizations in the world. It reaches more than 285 million people worldwide each month through its official journal, NATIONAL GEOGRAPHIC, and its four other magazines; the National Geographic Channel; television documentaries; radio programs; films; books; videos and DVDs; maps; and interactive media. National Geographic has funded more than 8,000 scientific research projects and supports an education program combating geographic illiteracy.

For more information, please call 1-800-NGS LINE (647-5463) or write to the following address:

National Geographic Society
1145 17th Street N.W.
Washington, D.C. 20036-4688 U.S.A.

Log on to nationalgeographic.com; AOL Keyword: NatGeo.